THIS RARE EARTH

JEREMY THOMAS GILMER

THIS RARE EARTH

Building the Dams, Mines
and Megaprojects that Run our World

Véhicule Press

Published with the generous assistance of the Canada Council for the Arts and the Canada Book Fund of the Department of Canadian Heritage.

 Canada Council Conseil des arts
for the Arts du Canada

Cover design: David Drummond
Frontispiece photo: Jumaring up the dam face at Toromocho, Peru
Photo on p. 206: Sun halo above a crane in Colombia
Typeset in Adobe Minion and Filosofia
Printed by Marquis Imprimeur

Copyright © Jeremy Thomas Gilmer 2025
Dépôt légal, Library and Archives Canada and
Bibliothèque nationale du Québec, second quarter 2025

Library and Archives Canada Cataloguing in Publication

Title: This rare earth : building the dams, mines and megaprojects that run our world / Jeremy Thomas Gilmer.
Names: Gilmer, Jeremy Thomas, author.
Identifiers: Canadiana (print) 20250187841 | Canadiana (ebook) 20250191946 | ISBN 9781550656794 (SOFTCOVER) | ISBN 9781550656848 (EPUB)
Subjects: LCSH: Gilmer, Jeremy Thomas. | LCSH: Mineral industries—Social aspects. | LCSH: Mineral industries—Environmental aspects. | LCSH: Mines and mineral resources—Social aspects.
Classification: LCC HD9506.A2 G55 2025 | DDC 338.2—dc23

Published by Véhicule Press, Montréal, Québec, Canada

Distribution in Canada by LitDistCo
www.litdistco.ca

For my Dad. You are so missed.

For Jenny Baird. Your love is what makes home.

Contents

INTRODUCTION 11

The Drop Zone
Toromocho, Peru, 2014 17

Kings Mountain
North Carolina, 2018 30

"Perigo MINAS!"
Soyo, Angola, 2005 48

Winter Steel
Rainy River, Ontario, 2017 61

Prayer Lamps for Lost Souls
Santa Cruz de la Sierra, Bolivia, 2012 73

Lake of Diamonds
Lac de Gras, Northwest Territories, 2007 87

The Price of Birdsong
Brokopondo, Suriname, 2009 104

How to Fix a Mountain
Ancash, Peru, 2006 120

The Strongest Winds in the World
Puerto San Julian, Patagonia, 2012 124

Maps of Old
Katanga Province, Democratic Republic
of Congo, 2010 137

Tunnel Between Worlds
Antamina, Peru, 2000 156

How to Kill a River
Lima, Peru, 2024 166

Checkpoint Lives
Antioquia, Colombia 2019 176

Jenga
Puerto Balboa, Panama, 2002 188

Mitigation
Minas Gerais, Brazil, 2025 195

ACKNOWLEDGEMENTS 203

Introduction

A STACK OF SMOKE rose into the canopy of the dawn as it broke over the horizon, the reds from smoke and sun mixing like a cocktail. I was driving out of Kamloops, British Columbia in 2022, along a wide highway cut through the dry, pine-covered hills toward a huge mining property. Next to me in the car was Don Hickson, an engineer, friend, and co-worker, who had just flown in from Peru. We were going to carry out an on-site inspection of a tailings dam tucked into the mountains ahead, near where those fires burned.

My phone lit up. Don answered it, and I could hear the trouble in the voice on the other end. It was an engineer in Brazil, and he was sending me an email. I pulled over at the top of a rise and looked at the small screen. There was a video, taken a few hours before. The pitch black of night on the frame was lit up with the shimmering hues of a fire, a burn stretching from the tree line and across the face of a dam. Another fire, another mine, on the other side of the world. The engineer was asking if they could hit the fire with high pressure water, which may have seemed like the right thing but in the world of tailings dams and high-risk structures, was not. I explained my objections and told him I would be passing on the message to our technical team in Toronto and Los Angeles and asked them to hold tight. I sent off the message, which was picked up immediately.

I sat for a moment, watching the fire burn on the phone, then looked up to see the smoke filling the sky in front of me. It was in that car, watching those adjacent events unfold simultaneously on two continents, that the shape of this book came into view.

This book is a graphic account of my twenty-five years working for some of the largest mining and engineering companies in the world. I jet all over the planet, often in the grip of some task: building a dam, drilling a hole, excavating earth, or raising a structure against the reach of gravity and odds to build something that may yet stand in a thousand years.

Much of this work was conducted in South America and Africa, often in conflict zones, on sites with heavy military and police presence. The co-pilots to these efforts were nearly always risk and excitement, wonder and horror. I was also not prepared for how those events would alter my view of the world, and our place in it. I have watched companies literally move mountains but have also seen communities dump poison into the rivers that ran past those mountain ranges. I have watched millions of tonnes of rock being extracted to make the materials the world uses, then watched anger generated in the populations that depended on those same materials. I watched companies lay waste to ecosystems but also fight to protect them. None of it was what I expected; the only constant was surprise.

I learned early to keep my eyes open and my ear to the ground, so as not to miss a stunning sunrise over Patagonia, or songs coming across the soft night air in Angola or catch the flash of some unknown creature disappearing back into the trees of Brokopondo. I have watched sandstorms on the Bolivian highlands and lightning over Congo. But I also have been robbed in a protest in Lima, watched police thwart an armoured car robbery in Santiago, and seen children walk through mine fields in Angola. The sum of all these moments, the camaraderie of colleagues I shared them with, and the struggle of completing those sometimes-gigantic projects, is what ultimately brought me to writing these stories.

I wanted to better understand what it was I was doing, where these efforts were taking us, but also what our work meant to the world and the tug of war between these things. The difficulty was not only deciding what to share, but how to capture the confusion of bearing witness to things that did not match the narrative of what I was hearing in the world; a world in environmental collapse, but also a world with a very poor grasp of the systems it relies on. Always in the background were questions: How do we depend on these activities, while making things better? How do we heal our planet while also taking what we need?

In *This Rare Earth*, I hope to pass along some of those questions—questions for which I don't have answers—in the hopes others may be asking them too and perhaps begin a dialogue that can connect us. I write from a place rarely heard from in the debate: an extraction industry with deep connections not only to the environmental challenges we face as a species, but also the very systems that support society and our everyday lives. I think readers will be surprised not by the obvious damage done by that industry, but by the reality of many of the complex decisions made by people trying to balance engineering and environmental conditions, politics and public opinion, and human demands upon the fragile ecosystems we all depend on. It's not unusual to find myself fighting for better methods, deeper studies, and more constructive decision-making in contrast to a show-me-the-money attitude from local and national officials. From development groups who want that gold at all costs to mining teams who spend years studying and planning for the protection of the environment, the world I inhabit is a maze of contradictions.

I hope these stories touch you, amuse and infuriate. Most of all, I hope they leave you wanting to know more about these places, the people who work in them, and the systems we are all part of, often without realizing it.

Jeremy Thomas Gilmer
Saint John, 2025

The Drop Zone
Toromocho, Peru, 2014

I AM STANDING ON a dam. It's a big one, around eighty million cubic metres, larger than some Canadian towns. The structure blocks the end of a valley that sits below a cluster of mountain tops, high in the Peruvian Andes. The lower slopes are covered in green-and-brown grasses, giving way to the crumbling rock of the peaks as they reach into a cloud-filled blue.

I'm atop this dam because I'm preparing to step off it. I'm in a harness, and beside me are climbing ropes threaded through carabiners—D-shaped mountaineering clips—and anchored to slings cinched tight around my Hilux pickup. I'm flanked by rescue personnel tasked with monitoring me, inspecting my descent and ascent system, and staying close in case of an accident or some other complication. A gaggle of bright red jumpsuits and helmets, they look like a Devo tribute band and are easy to spot. Among them is Manuel, one of my field staff and a seasoned operator in dam construction. He looks over my workspace for anything out of place or in the way. He gives me a wink and a thumbs-up.

As the construction quality assurance manager, I am looking for cracks or deformations on the dam's concrete upstream face. I will inspect every inch of it, moving the truck every ten metres to reach each new section. I've given the rescue team two instructions. One, if I pass out from altitude (the ascent is a physical hardship at 4700 metres), the backup line, rigged to the back of my shoulder harness, is there to pull me up. Use it. And two, no one touches the rope, or enters the safety zone we have cordoned off. No one, but emergency personnel who I vetted and have full confidence in. As a further precaution, the truck keys have been carefully stashed in the truck, and an orange cone has been placed around the truck, including in the driver's seat—preventing someone from driving away.

Here's where the story gets a little more interesting. Strictly speaking, I'm not on a dam as much on the edge of a crest, or the angled slope of the dam face holding back a lake of blindingly blue water and soft reddish-brown material. This body of water and material is called the tailings. It fills the entire valley from one slope to the other and carries on as far as the eye can see. Small rivers flow across the surface, creating landscapes of colour. It looks almost like sand.

The tailings are the by-product of processes taking place over the peaks to the west of me. Sitting like game pieces on a city-sized chess board are ten-story, metal-clad buildings attached by a maze of conveyors and roads. Inside these structures, freshly mined and crushed ore is being turned into copper. But the process also churns out another thing: waste. Running along the edge of the mountain, beside an access road cut into the sides of the valley, is a large, steel pipeline that carries this substance—tailings—out to deposition points. From there, the liquid blasts out of open spigots, pipe sections that split from the main line and point down into the lake below. The release is deafening. Standing nearby, you can feel the thunder in the ground as the fluid rushes through the pipes, which are a metre in diameter. But down here, on the crest of the dam, I am only faintly aware of the sound.

I have spent years working at this copper mine called Toromocho, named after a bull-like mountain peak which was mostly removed during the initial phases of development. The higher peaks in the ridge that borders us reach over 5,000 metres in elevation. I was here from the early days of the dam, preparing the massive foundation for the whole structure. Excavators spent months digging out the topsoil and then digging down past the deep and wet organic layers to the hard bedrock. Dozens of dump trucks, tires as high as minivans, would climb out of the excavations, hauling loads of earth up and out on the impossibly steep access roads to stockpile areas where the materials would be stored until they could be used for something else. Often, they would be left in a huge pile, an artificial mountain of its own.

Eventually, the rock face showed itself: a winding complex bed of stone undulating like an ocean wave frozen in time. Geologists pored over it, mapping and recording the geological structures that lay there. They looked for softer rock that could pose a problem, or artesian wells: pressurized underground springs that could surge through cracks and complicate construction above. We would walk the ground, our team of engineers, scientists, and technicians, mapping the kaleidoscope of the surface, the colours and shapes telling a story hundreds of millions of years old.

As I adjust my helmet and glasses, I glance the other way, downstream in the direction of how the tailings would flow if there was no dam. The long drop of the wall goes for hundreds of metres, narrowing as it peters out toward the toe of the slope. I see the mining camp far below, a series of three-story stacked containers, where thousands of staff live. The camp resembles what I imagine a Mars colony might look like; thousands of people living by measured quantities; litres of water, volts of electricity, kilos of food. Everything measured but the air, and even that is in short supply up here.

It is go-time. I check and recheck the lines, communicate with the rescue team that I am descending, and I set my weight against the lines and walk back over the edge of the dam. My boots loosen some gravel, and it scatters toward the waterline below. I watch the stones fall and am reminded of the height. I rappel down a few metres and pull my rope tight against the figure eight, a pale-blue aluminum device used by climbers to slow their descent.

The dam face is called a concrete curb. I'm looking for signs that something went wrong in its installation, that things are not as they should be. The dam was built in stages. Rockfill was hauled in and dumped in layers, which were then cut and compacted using graders and bulldozers, equipment as tall as houses. The rock sat in piles as other machines spread it out in carefully measured sections, called lifts, and levelled it to ensure the larger chunks didn't bunch up and segregate. Once the rockfill was placed and tested, crews began placing smaller materials, gravel and sand, in strips across the outside edge of the rock on the "upstream side" facing the tailings lake. These are filters that allow water to pass through but hold back any coarse-grained bits of the tailings. The filters are placed in series of ever finer layers to prevent anything that slipped through from moving, or migrating, past the layer behind it. Finally, crews used a curb machine that extrudes a triangle-shaped piece of concrete, almost like toothpaste, along the edge of the dam, about a metre high. This curb rises, row after row, creating the forty-five-degree angle of the dam—the very wall I'm standing on. As the damn fills, the water and tailings rest against the concrete.

The process is undertaken by hundreds of people and dozens of machines, over the years. At night, the worksite glows: light from plant towers shine down, gathering swarms of flying insects; the spinning roof lights of pickup trucks and support vehicles; headlights of all kinds; flashlights carried by workers, their bright reflective vests and glowing shoulder beacons appearing as small floating sparks from above. The constant drone of backup alarms on equipment fills the night air as well, horns to signal movement or intent. And the never-ending radio chatter, voices in Spanish and English crackling over the speakers as information is passed, and questions answered.

So here I am, in the bright light of day, hanging over the edge, my eyes tracing a path across the small section of the dam I've assigned myself. I look back and forth across a five-metre-wide area to find and map any issues; patches of poor-quality concrete, spaces between sections that look larger than they should be, signs of damage from falling rocks, or unnoticed damage that occurred during construction. My boots are set against the wall, and I am leaning back in the harness. I step gently from side to side to stretch my vantage as I look over to the far edge of my zone of interest. Nothing noted. I drop down three metres and repeat the process. Sometimes I spot something, maybe a small crack, and taking my notebook from my vest pocket, I describe what I see, marking the location by counting the number of curbs from the crest of the dam to where I am hanging. It is a bit rudimentary, but others will be able to locate anything with the information I am recording, and that is the point.

I take a moment and look out over to the abutment, where the dam meets the mountain. The exposed rock is beautiful—tiered geological formations of varying colour and size. It is like a history of the region but written in shapes and not letters. To be up high is a gift. I spent a good part of the 1990s rock climbing in Western Canada, so a harness is familiar ground. I breathe in the thin air, I let the sun wash over me. Compared to my usual day in Toromocho—up before dawn, driving hundreds of kilometres over rough mine roads, managing a staff of dozens of engineers and technicians and support staff, constant meetings and field inspections, relentless problem solving—this is a mental vacation.

I have been neck-deep in mining since 2000, although the work started before that. I took a job in the summer of 1994 as a technician in a soil laboratory my dad ran in Burnaby, British Columbia. The lab was down some staff, and the work was pouring in. It seemed a good fit. I needed money for school, and I had hung around his colleagues growing up.

The arc of learning, the mix of physical labour, and exposure to engineers and scientists had its effect. Notice was taken of my work ethic and enthusiasm. I found the tasks interesting, and the staff was engaging and patient: lab techs from Japan, geophysicists from South Africa, and geologists from the UK and Chile. For a young man in search of direction, there was a lot to do and learn.

Over the months, my tasks became more complex. What began as a few excursions to retrieve samples became a fall and winter of field tests. I was dropped like a toddler into a playground. I drove all over the lower mainland of Vancouver to job sites and excavations. I spent days in the pouring rain monitoring drill rigs, only to arrive back at the lab to notes left by staff long departed for the day: "Could you get these two hundred samples in the oven? Will these Atterberg Limits be completed by the weekend?" I spent my nights grinding through tests and results processing. But it wasn't a chore. It was thrilling. I learned how engineers used those results to understand soil conditions and then designed solutions to avoid issues during construction. As the two worlds, lab and field, began to mesh, I saw through an even longer lens: the decades of study that went into the science of how we understand the world around us and under our feet.

I still dreamt of being a novelist, a writer of many things, and I even did some magazine work. But the emails kept coming; the engineers from the office above kept dropping by. "Jeremy, we have a bit of work in Alberta/on the island/ out east. Would you be interested?"

Looking back, I said yes to nearly every opportunity, every project. I began gathering momentum, my name was thrown out whenever someone was needed, on short notice.

Then one day in 1999, I had an email from Terry Eldridge, a very senior and respected engineer, a fellow I had known through my father for years. Terry worked for the mining group. They were the big boys, responsible for projects in far corners of the world where things like copper, diamonds, and gold were extracted. Projects where teams of experts and professionals were sent to solve complex problems under tough conditions. Projects that pulled you in for months and sent you home with a tan and a bag of stories. Projects where careers were built.

Terry called me into his office—his wall and desk were covered with drawings and technical images of all kinds of structures, dams, and open pits. He waved me in, and I sat while he finished an email. He turned to me and smiled, his huge, Klondike moustache framing his large face. He had always been good to me, treated me with a level of respect, and always seemed interested in what I was working on. He was also someone not to disappoint.

"So, we have a job in Peru, up at a copper mine we have been working at, and I think you would be a good candidate for supervising the construction of a dam up there. It is not much different than the work you have been doing for us, but it is a more challenging environment. I think you would really like it, and you would be a good choice to go do it."

The blood drained from my face. I was both terrified and excited. This was a vote of confidence I had not yet experienced in my work life. I was smart, and people liked having me on their jobs, but this was different. I instantly understood the scale of the responsibility, and the care it would demand every minute of every day. I looked up. Terry was still waiting for my reaction.

I smiled, probably levitating off the chair.

"I'm in."

I began to ascend, having inspected the vertical wall of this section of the dam. I take two jumars from my harness belt. These are large metal handles that can move freely up a rope onto which they are clipped but lock when downward pressure is applied. I am holding jumars in both hands as I grind up the face, step by step, the sound of my feet and the jumars whizzing up the rope create the rhythm of my ascent. As I rise, I keep an eye out for anything I may have missed on the way down. The climb is exhausting at this altitude, my lungs are bursting for air, and at times it feels like drowning. My muscles are now suffering from the lack of oxygen as well, with cramps in my biceps and thighs. I look up now and then, to see the crest inching closer. I find that the best way is not to stop and rest but to push through.

The red hats look down from above, lots of smiles and thumbs-up. They know this is the tough part; they can see the pain on my face. They are well equipped to fetch me in any kind of emergency. Over the next days, their numbers will dwindle, actual emergencies and training sessions will pull them away, but there will always be two with me, ready to go and happy for the activity: descend and ascend, descend and ascend.

I reach the top and get well clear of the edge before I stop, in case I pass out and tumble over. After a few minutes, I gather the ropes, review my notes, and move the truck another ten metres to the next section. I had budgeted just over two weeks to inspect the entire dam.

My record was sixteen descents, and sections inspected, in one shift. The pace is crushing, leaving me a little sorer, and a little more tired. Just another ten days to go. Eight days. Six days. Almost there.

So far, I've mapped only a few small cracks, nothing that would denote structural issues. My left elbow, my lead arm while ascending, developed an ache I had not experienced since my younger days of climbing granite in Squamish, BC. The wear and tear was taking its toll.

I had gathered my ropes and moved my truck to the next location. I had one more run left, and that was it for the day. The sun was about to fall below the mountains, and I could already see the long shadow crossing the valley. I looked far up on the left valley walls and spotted bright yellow wreckage. It was all that remained of a huge D-10 dozer that had toppled there nearly a year before. I watched the setting sun reflect off broken glass and steel.

It happened on a night shift. A young operator from one of the local towns in Ancash had been bulldozing freshly blasted rock over to a loading zone where an excavator could put it into the haul trucks. He was working on the flat summit which had been drilled for fill—material needed for the dam and other site projects. Moving mountains from one place to another was a task often undertaken on operations of this scale. I was always awed by the scope of it but also torn at the fact that we were changing the very skyline of a valley. It was something done by man a million times in a million places, but bearing witness to the act is very different than sitting through a history lesson; to hear and smell the blasting, see the trucks hauling rock, watch the top of a mountain lower every few months. I thought of pyramids and ancient cities built on the Nile, I thought of railroads cut across the lands of Kenya and the American West. It was the act of winners, and there were always losers.

The operator was surrounded by steep drop-offs, and when that's the case, the rule is to build a high berm—a long ridge of material that serves as a wall—to keep vehicles from going over the edge. For whatever reason, the berm was not there; or maybe it was too small a berm, one that was no match for such a large machine. In any case, sometime during the night, while talking on his cell phone, the operator backed up too far. The dozer dropped about 300 metres. The young man did not survive.

There are many rules regarding cell phone use in work zones, especially while operating equipment. At every step of training, not just the specialized kind an operator receives, but also the visitor induction required to set foot on site, warnings are repeated, and repeated, and repeated, and repeated. But habits, it seems, don't change. And now, all of us on the dam crew had a monument, a lasting reminder of the cost of bad decisions.

I am jumaring back up the slope again, the last one on this round. I have just a few more days of this, and then I am done. My gloves are burned through. I hope they last the effort, though some duct tape will see to that. I stop at a small crack. Mark it, measure it for depth and length, locate it on my map, take a quick photo, and keep moving. The wind is starting to kick up a bit, dust is blowing off the face. I look back across the lake of tailings. I see the flashing lights of trucks, high up the mountains.

I get to the top, get clear of the edge and unceremoniously undo and drop my waist harness. Stepping out of it, I stretch, and the rescue guys are already helping pull up the lines and packing our gear for tomorrow. My hands have trouble gripping the steering wheel when I drive away. Crews are working the dam again; a grader will begin cutting down sand for the next lift now that I am off the dam. I see the night shift buses rolling in, blue hard hats staring out foggy windows.

Passing over the dam, I look once again at the folded geology of the rock, a monolith jutting over the landscape. The camp is not much further on, a warm dinner and a bed waiting. I turn up the music, the coppery taste of altitude still in my mouth.

Kings Mountain
North Carolina, 2018

The two small boys were swinging their feet as they sat in oversized rocking chairs on the sidewalk leading into the mall. They were laughing amongst themselves and then quieted as I approached, watching me walk up the parking lot. They stared, and they smiled.

"Hello gentlemen," I said.

"Hello!" they chorused. One of the boys said something I couldn't hear.

"What was that?" I said.

"Where are you from? You talk funny." It was the smaller of the two. I assumed they were brothers due to their identical smiles. The older one smacked the younger playfully on the arm, and I wondered for a moment if he did that because I was a stranger or because I was white.

"I'm from Canada, I live in a little place called New Brunswick, about a twenty-hour drive north from here. What are you guys doing on this fine day?" I asked.

"We're waiting for our mama, she's inside." The older boy said. The younger boy was staring intently at my face, perhaps looking for some sign or hint of my Canadian foreignness.

"Well, it was very nice to meet you boys, my name is Jeremy," I offered.

"My name is Christopher, and this is my brother Marcus," he said, sitting up a little straighter, his voice a little deeper, his manners taking hold.

"Well, I am very pleased to meet you. I hope you are all set for the next storm," I said.

"Yes sir, we are ready for everything," he said.

As I walked into the mall, the boys returned to swinging. Further out, the parking lot was filled with what could have been 150 hydro trucks, complete with lift buckets and painted logos from several states and provinces. The greatest surprise was seeing my province in the mix, boxed among the others, the workers having crossed the street to one of the numerous hotels and restaurants. I doubted anyone back home, save the linemen's families, even knew they were down here, helping the area recover from one of the worst hurricane seasons in history. And it was just getting started.

Kings Mountain is about an hour west of Charlotte, but the community is a world away from the sports stadiums and mansions of the larger city. Just outside the town sits one of the world's leading research centres for lithium production. Over the past two decades, the use of the soft, silvery-white metal in lithium-ion batteries—a key technology in electric cars—has skyrocketed. The facility also sits on a closed lithium mine from the eighties. As part of a potential expansion, a drilling program was launched to assess if conditions were favourable to begin mining again. The work was conducted by a small outfit out of New Brunswick and used a mix of English- and French-speaking drillers. It may have seemed strange that the company had been brought in from so far, but I had seen Canadian drillers on the ground in Peru, North Africa, and elsewhere, and they were among the very best in the world.

Hanging over the entire operation, however, was the fact that it was about to get rattled by some serious storms. Which is why I was here. They wanted someone with drilling and safety experience to help wrangle the teams should the worst come to pass.

I take hurricanes seriously. In 2005, I was helping run a project in Angola when Hurricane Katrina tore through the Gulf Coast. It left behind not only apocalyptic damage and hundreds of deaths but also exposed the shortcomings of a powerful nation that showed itself politically and technically ill-prepared for such an event. I could remember watching my friend, a geotechnical engineer from Houston, arguing with his wife over the phone about what to pack while she evacuated their house, the fear evident in his face. Hurricanes could end the world around you in a matter of hours, sweeping away the support systems you depend on like water-logged debris.

My official title in Kings Mountain was project safety manager, which meant the lives of a dozen geologists and the full crews of six drill rigs were squarely in my hands. If there's a risk on-site, my team's job is to uncover it and fix it. The current project involved many moving parts, with staff dispersed across various drill rigs. Some rigs were deep in the dense North Carolina forest, reachable by small four-wheel buggies. Additionally, two drill rigs floated on barges in the lake covering the pit, accessible exclusively by small boats.

Mornings started with my night-shift staff in our site office. The night shift covered the always-active drill rigs. Most mines operate around the clock to keep the project on schedule. My safety manager was a wily and fit man in his early fifties I'll call Bill. He had spent a career in the US Army as a specialist in Explosives Ordnance Disposal. This meant he worked with soldiers in identifying and clearing bombs, shells, or landmines. Bill had trained soldiers all over the world and had risen to a role in the protective detail of a certain vice president. His stories were the stuff of legend—like the time he defused a roadside bomb under sniper fire, sweat drenching his protective gear. Bill was an exceptional safety guy. He could spot a bottle cap in the dirt as you sped by at 60 kilometres an hour. He was also funny as hell.

Over coffee, he gave me the skinny on the night's progress and whatever safety issues the drill crew had. He always saved a bitching session for the end ("I put an order in for chainsaws for all the crews, provided they don't go cutting their own feet off, stupid bastards"). These mornings were just the kickstart I needed.

The site was spread out over kilometres of old-growth forest. It was primeval: tall Virginia pines and American Sweetgum climbing out of the thick cover of vines and flowers. I drove my side-by-side—a small, open, four-wheeler. The roads could handle trucks, but the wear and tear on the tracks, plus the costs involved, made using these agile vehicles more sensible. I headed down what looked like a dead-end track and could hear the drill rig before I saw it. A large box on skids with a boom reaching into the air at a sight angle. Core boxes and other gear lay on the ground in the clearing that had been cut out of the trees.

The sun was already up and cut fine beams of orange and yellow through the hanging mist. Floating dust moved along invisible currents. I took a moment before going into the work area. Birdsong spilled from the trees, either undeterred by the drill's engine or answering it in kind. One of the driller helpers came out of the box carrying a wrench. He waved before grabbing a long pipe off the rack and heading back inside. I poked my head inside the door, got a nod from the driller, and watched for a few minutes. The area was neat and organized. I walked back out into the pad and watched from outside. The closure around the rig cut the noise down substantially, and I breathed in the air flowing through the pine trees wet with the dampness of morning.

I descended the steep road down toward the pit, light bouncing off the surface of a man-made lake deep enough to submerge a hotel high-rise. I could see one of the two barges, its deck supporting a drill rig, and a small boat moored beside it. Long steel cables held the platform in place, anchored to points along the pit's walls and shoreline. Strung along each line were small plastic flags, fluttering where the warm air lifted from the depths, riding up the sheer walls before catching the westward breeze.

I pulled up to a small dock, where an aluminum boat with an outboard motor bobbed in the mist. I strapped on a life vest and climbed in. This was always the best part of the day, which is why I made it my first stop. The engine kicked over, and I steered toward the first rig. At the access dock, I tied off and stepped onto the platform, stopping at the safety line. I might've been the safety manager, but it was still the driller's rig—and you waited for permission. I spent half an hour inspecting the rig and talking to the driller and his helper. The cold blue sky was giving way to fast-moving clouds. Around the rig sat piles of steel pipes, a drill mast, and metal drums full of fluids and water for the drilling. I imagined what hurricane-force winds would do here, raging off the rock walls and tossing gear and people into the water. I imagined the sounds that the steel cables holding the rig in place would make when tasked beyond their limit, the sharp crack as their anchors broke free.

Soon I was back in the boat, speeding across the water. The other rigs were deeper in the forest, down narrow paths of mud, and cut into tree lines. I would visit them one by one, check in with the teams, note what depth they were drilling at, look at core boxes full of soil and rock, and check for safety issues or improvements to make. As I drove around the site, the skies continued to change.

Back at our our main office, I walked into the "tents": a part of our operations hub sheltered by tarp-like roofing and curtained with thick plastic sheets. Here, dozens of boxes filled with rock core sat on long, wheeled rails, designed to slide easily from one station to the next. The geologists were logging and recording what the drilling crews had brought them: iron and feldspar, phyllites and schist. The team took photos, made notes, and discussed endlessly. Lithium, the reason we were all here, was nearly never mentioned, as the underlying geology was far more interesting. The group was a mix of Canadian and American geologists. They worked under a manager I will call Ryan. Ryan was a leader who did not always wear the role well, often losing the big picture in the minutia of the day-to-day. The real rock star was Chris, the intellectual heart of the team. All questions eventually found their way to him, and his joy in the work was infectious.

"So," he said. "How are the storms?"

Yes, the storms. Two large tropical formations had begun forming over the Caribbean and were pointing straight at us. The client made it clear—they needed me to evacuate the ground team in case of a hurricane. In a prior storm, Ryan had resisted leaving, and they wanted someone who could take charge and ensure it didn't happen again. I understood.

"Yeah, no improvement there; we are looking at a few more days, maybe less and then we may have to clear out. What is your timeline for getting all outside core back in the containers?" I asked. Chris sighed.

"Man, Ryan isn't going to like that," he said.

"Then good thing it isn't up to him," I said. Chris grinned at that.

"They are calling it Florence," I said to the room, referring to the storm. The whole geology team and several of the group leaders for the drilling outfit were huddled there. The safety manager for the mining company stood next to me. His name was also Jeremy, and he was from Arkansas. Jeremy and I differed heatedly on all things political ("Well, that sounds like some coastal elite talk!!") but saw eye to eye on safety. As I began to lay out the shutdown protocol, evacuation planning, and storm sheltering plans, I could see Ryan fume. His face was getting angrier with each passing second. He had a team on the ground and thousands of hours of work ahead. They had rock core coming in every few hours and were under stress to get it properly logged and reported in a short amount of time. I sympathized with his workload and the seven different directions he was being pulled, but I also knew whose decision this would be, and how the decision was being made.

The coast guard had been pulling boats off the water and redirecting ships out to the sea for nearly a week. All over town we could see homes and businesses boarding up windows and preparing for the storm. Jeremy covered the mine's preparations and the community's response. He noted that local rescue and fire had committed to prioritizing the project in the event of massive damage or impact from the storm. I knew this would be a point that Ryan would bring up, and I knew why it was a problem. We finished the meeting, and, as I expected, Ryan came over as the room was clearing.

"This is bullshit," he said. "We don't even know if we are going to be seriously affected, the storm may blow right over or fizzle out. And you heard what Jeremy just said, if there is an emergency, we get priority for rescue response. So why are we packing it up man? You scared of a little storm?"

His response did not surprise me. It was the response of someone who had spent a lot of time getting projects done on time and budget. It was also the response of someone who had never actually been in harm's way. Ryan was very good at what he did—safety was not one of those things.

"Ryan, geology is your wheelhouse, you run a tight ship, and I respect that. Safety is mine. It is the reason I flew here to manage this. It is my decision, and we are going to wrap up operations in the next thirty-six hours and button down everything for this storm. My job here is not to convince you of this. It just is."

"Well, we will see about that. What are you gonna say when the storm blows over and nothing happens?" he said.

"I am going to say we were very lucky, and let's get back to work," I said.

"Okay, so what about our rescue priority. Let's say we did get hit a bit, but we have people who can come sort it out for us."

"Ryan, there are tens of thousands of people in this community, and over ten million people in this state. Most of these people have nowhere to run to. They are going to ride it out. We have the option of shutting down and getting shelter or leaving. In a catastrophic storm, there could be massive damage, loss of life, and people stranded or in need of help. If that happens, I don't want local fire and rescue teams fucking around with our guys, who had every opportunity to move out of the way. In fact, if this thing hits hard, I fully expect my guys and myself to be assisting in those efforts. We are all first aid trained, some of the drillers have rescue experience. If this goes sideways, we will do what we need to. That won't be drilling. Does that make sense to you?" I said. I was doing my best not to get upset. I was failing.

"Whatever, whatever," he said.

I looked at Ryan, seeing the failure of connection.

"That right there?" I said, "That is why this is my decision and not yours."

Thirty minutes later, geologists began packing boxes of core and stacking them into the steel shipping containers that sat out on the laydown pad. I thought about cars I had seen on the roads, roofs loaded with whatever families could tie down. The phrase "climate refugees" had lingered in my mind over the past two weeks, but witnessing it come to life in Kings Mountain was jarring. It was a term I'd encountered for years: in the meetings I attended, in the books and stories I read. But here we were, not fiction, not film, real people leaving their homes behind in the face of calamity.

Two days later came the winds, followed by flooding. The drill rigs had been battened down, the floating barges were moved to shelter in the far corner of the pit, out from the center but not so close that they could bump into the walls or each other in high winds. Everyone was instructed to fully charge phones and radios and fully charge backup batteries in the event we lost power for days. Buckets of water were stored in bathrooms so that toilets could run if the water was shut off. Most of the geology team lived in several rented homes in Gastonia, half an hour away. My drill team was with me, housed in some Holiday Inn closer to the site. Our hotel was on high ground and close to the highway. As the winds began to kick up, and news of coastal damage began filtering in, I was relieved that I had no people out in the field. We hoped that as the storm transited overland, it would lose steam.

The wind blew hot now, an electricity in the air that seemed to make everything hum. Jeremy and I were doing our tours of the project. The drilling project was only a small part of the operation. Not far from headquarters was the actual lithium plant. The plant had its own set of issues. Lithium production cannot just be switched off like a light. There were systems that had to be kept running under almost any circumstances, lest fire and possible explosions occurred, and the company had left a skeleton crew in place to oversee this. I was one of only a handful of people with permission to be on site now. There was also a heavier-than-usual security presence to ward off locals who might think it was a good time to grab whatever they could find.

Later in the night, I was sitting in my truck talking to Aaron, a plant guy tasked with monitoring the south end of the plant where chemical and material storage filled two large warehouses. He sat in his own vehicle, window open, and yelled across that the news had linked deaths along the coast to the flooding.

"People don't wanna leave their homes," he said between cigarette pulls. "I get that."

"Yeah, sometimes those people are geologists," I said.

Aaron laughed, fully aware of the challenges I'd faced getting Ryan to clear out his team. I think what really bothered Ryan was having the decision taken out of his hands. As he and his crew moved on to bigger and more ambitious projects, the importance of safety was a lesson he'd have to learn again and again.

A steel barrel lid came clanging down the pad and dashed past us. As the gusts grew stronger, metal panels on the sides of the buildings began to vibrate and sing. The wind spun in tight circles over the surface of the asphalt, carrying off loose gravel and bits of leaves like Jinns escaping from a bottle. Flecks of water, not quite rain, smacked against my car and struck my face through the open window.

"Well, I am gonna do a quick run around by the office, see if anything is going on there. There is a small crew in the labs, checking on things and ready in case we lose power. I expect that we will any time now. The backup generators have about four days of fuel, but we gotta keep them topped up. I will see you back here in an hour," he said.

I drove up to the end of the paved road where our drilling complex was, then took a circuitous route around the whole property. I noticed that even though we did not have heavy rain, the drainage ditches were quickly filling, and that water was beginning to pound on the surface of the parking lot. I could feel the earth seething below me. I thought about those on the coast, streets under four feet of dirty water, roofs blowing off, and windows smashing in under the raging storm. I thought of thousands of people crowded into school gymnasiums and community centers, and of those who could not make it there, huddled in station wagons or on kitchen floors, unsure of what to do next.

My phone lit up, a message from my boss in Canada, Mike, checking in on things. I wrote back that the full storm had not hit us yet, but we were feeling it now. Mike reminded me that the client had asked me down there to help support any evacuation, and that included pushing back against resistance from the contractors.

I returned to the main office compound and found

Jeremy and Aaron talking in the parking lot. I pulled up and said I was heading back to the hotel for a few hours sleep. We still had six or seven more hours before the full force of the storm was on us and there wouldn't be much chance to rest after that. Jeremy agreed, saying he was going to sleep in his office. We would all check in at 4 a.m. Jeremy would message if there were any problems.

Arriving at the hotel, the clouds were black-and-grey bubbles hanging low. I found a family out in front. It was a woman, a young girl (who I assumed was her daughter), and a grandmother in a wheelchair with an oxygen tank strapped to the back. I said hello as I joined them to watch the sky.

"Good evening," said the grandmother, as all three nodded. "It is some sky out here, telling us to cover our heads." Her language felt like something from another time, an older English.

"Indeed, it seems things are quite bad on the coast. I am hoping it blows out a bit and loses some of its strength on the way in," I said.

The grandmother nodded, taking a long haul on her oxygen line, a deep rasp coming before the words. "These storms, they are brave over the water, but they lose their courage once they get overland," she said.

The family asked where I was from. I told them and said that I was working at the lithium company (they all knew it, being one of the larger employers in the area), and had evacuated my people, some of whom were shacked up here at the hotel. The grandmother, whose name was Betsy, told me they lived in a small house in the town, which sat on low ground and was already flooding. They had been sent to the hotel. She said they were very lucky, as many folks had been sent to local shelters, but her health had put them on the list for this place. I told them I was happy they were here, high and dry. As I left, Betsy called me back and took my hand in her fragile grip.

"You take good care of your boys, I will be praying on them," she said.

The call came in at around 11 p.m.: all staff were barred from the project site. I called the geologists in Gastonia and gave them the heads up, just to be sure no one did anything stupid, and confirmed everyone was buttoned down. The drillers were in their rooms, hoping the TV didn't go out.

We were locked down for two days. The coast was battered—thousands of homes were flooded and damaged by winds. It was called a Cape Verde hurricane because the geography of its genesis was deep in the South Atlantic over the Cape Verde islands off the coast of Africa. Florence had reached Category 5 by the time it struck the coast. Of the twenty-two direct fatalities, fifteen were in North Carolina. The storm was briefly downgraded to a tropical storm on September 9 but then gathered strength and returned to a Category 4, something rarely seen in storms.

Although we were in the path, we were very lucky.

Except for a few hours without power, and serious flooding around the hotel, we were unscathed. I watched from the third floor as emergency vehicles made passes up and down the highway, white power trucks racing out to communities at all hours, repairing lines as soon as there was a break in the wind.

The residents of Kings Mountain gathered themselves and got to work, cleaning up. But among the shovels and pumps, chainsaws, and generators, I saw a town sitting at the tail end of a meteorological corridor that shows no signs of relenting. Weather models from the National Oceanic and Atmospheric Administration paint a grim picture for the coming decades: endless and devastating storms. Following Florence, no fewer than twenty-seven tornadoes developed in places like Virginia, where tornadoes are not traditionally a source of fear. Stream runoffs and sudden phreatic shifts caused landslides and sediment damage that lasted for weeks. I thought about those dozens and dozens of white bucket trucks, from other states and provinces, who came when called. What would they do when storms strike a place like Kings Mountain on multiple fronts, all over the east? Who gets your power back on then? Who runs hospitals when a hurricane hits in the middle of the next pandemic?

The perfect storm isn't just about wind and water. It's the collision of unstable weather and an unprepared public; it's cascading environmental impacts and a distracted, misinformed response. What happens when governments that reject climate change direct money away from preparation and studies, and rail against teaching basic science in schools? What happens when organizations that protect water quality are sidelined, and deregulation spreads just as fresh water is more and more threatened by events that a portion of the population neither believes in nor understands? There is something surreal about driving through the path of a storm, hours or days after it has raged over the roofs and heads of citizens, only to see those same citizens raising yard signs proclaiming climate change a hoax.

I often brag about the talent I encounter in my in-dustry. But I also work with intelligent and thoughtful peers who seem to be politically and culturally immunized against certain concepts and information. I've sat through meetings where budgets for future climate impacts and responses were carefully debated, only to hear people joke afterward that "fantasy hour" was over. They may not represent the majority, but their presence still influences outcomes. Ryan chastised me after the storm because there wasn't enough destruction. Our evacuation, he said, had been unnecessary and set their schedule back. What happens when more people like him start making the decisions?

We returned to the site and began getting things back up and running. Weeks of sunny weather followed, and the resulting explosion of green made it feel like the land was breathing again, having held its breath during the storm.

But the mornings were cooling too. The kudzu was covered in an overnight trim of frost, putting a strange halo on the hanging carpet of leaves.

I stood above the pit, watching the crews guide the barges back into place so they could start drilling again. I watched the occasional cloud travel across the skyline and thought about the next storms, this year and next, this decade and next. I wondered what we needed to do to be ready. I am still wondering.

"Perigo MINAS!"
Soyo, Angola, 2005

MORNINGS IN SOYO were a rusty haze. Our small boat pulled away from the dock, but even this far from town, the stench was unmistakable: burning trash. Grey and black smoke billowed up from behind the walls surrounding the base. For me, this was the smell of poverty.

Our pilot guided us around listing steel ships. Perched on crow's nests and along the decks, birds watched over the remaining skeletons of the abandoned dead. The water was slick with a sheen of oil that leaked from these hulks, the riverbed clogged with bullets and unexploded shells. The ships were relics of a rebel group's desperate attempts to retake the port by force during the fading weeks of the civil war that ravaged Angola for over three decades. It was into this scarred land that, in 2002, gas and oil companies started pouring people and resources. I was here with a team of drillers and engineers working on a huge Liquefied Natural Gas (LNG) construction project. We were carrying out a drilling program, both on water and land, to develop a better understanding of the underlying ground conditions where things would be built. Along the shoreline, an LNG receiving facility would be erected, a vast city of pipes, tanks, and steel frameworks. Once complete, it would be visible from kilometres away or from thousands of feet above.

We were headed for a small drill rig—a barge the size of a motorhome, set up on stilts—to load core boxes of soil samples to take back to our lab. Out on the water, I could see people on motorized dinghies or dug-out canoes. People were known to float across the river from the Democratic Republic of Congo, on makeshift craft, seeking refuge from even greater turmoil and scarcity of opportunity in their homeland. I heard stories of lone children drifting on inflated inner tubes from truck tires. I imagined trying to cross this wide river, at night, as a boy, the blinking lights of the far shore my only source of direction.

The town of Soyo is on the northern tip of Angola and next to Kwanda Base, one of the largest oil and gas extraction projects in the world. A plateau of concrete, buildings, and machinery slapped onto the end of a river delta and marshlands stretching out of sight. Kwanda Base was home to a helicopter fleet and support centre that fed dozens of drill rigs with workers, and a navy's worth of ships. Freighters of all shapes and sizes shipped millions of tons of materials and equipment monthly to structures that sat, like alien atolls, off the coast of the country. These offshore platforms carried out the drilling and then pumped and extracted the oil, along pipelines or onto ships, for transport back to the base.

Although the civil war had been over for years, remnants of the conflict and the instability it left behind were still a daily challenge. Attacks on equipment convoys, kidnappings, and the occasional gun battle were normal. Regular sightings of "technicals"—heavy machine guns, or other such armament too heavy to be wielded on foot, mounted on pick-up trucks, often used by militias—left one with a sense of general chaos.

Camp life on the base, however, could lull one into a false sense of home: neat rows of furnished bungalows with yards and BBQs, steaks and beer in the fridge, satellite on large-screen TVs. Neighbours walked the length of their driveways to fetch a copy of the *Post* or the *New York Times*. But rise early enough and the illusion faded. The paper was not delivered by a boy on a bicycle but an Angolan security guard from the open box of a pick-up truck, the barrel of an AK-47 pointing out an open window.

Most days I was not out on the water; I was on the base, running the laboratory and helping organize the team. In the early mornings, I would drive my battered Range Rover into our area of operations, past office complexes and warehouses until bright signs warned me to stop. Orange-vested traffic controllers would hold us back as helicopters rolled past us on their wheels. The helicopters would line up to take off one at a time, a morning ballet of Super Pumas and Sikorskys full of men being ferried to offshore rigs, far out at sea. The constant departures gave the place the feel of an ongoing war, just over the horizon.

I pulled up to a warehouse at the edge of the base. A group of Angolan and Congolese workers stood nearby, staring at our foreign crew gathered for our morning meeting. The Angolans whispered to each other, taking in every motion, every word, then mimicked something that was said, some motion or body language. My guys paid no mind, used to the attention. They were all smoking except for the doctor. Rudy Vasquez received his PhD in geotechnical engineering from Berkeley. He would introduce himself to you with a firm handshake and "Rudy Vasquez, PhD." The first time I thought he was joking, but he was completely serious. The other guys loved Rudy but took the piss out of him relentlessly. The crew took medical com-plaints to him: *Rudy this itches, Rudy I can't stop hiccupping, Rudy I have explosive diarrhea!* He always replied with the same dry turn, "I'm not that kind of doctor."

After a morning brief, done around the hood of a truck covered in coffee cups, we would be joined by our security team, two middle-aged men assigned to our group. Bert and Paul were friendly, terribly fit, confident, and intelligent. Former South African special forces, they were from the now disbanded 38 battalion and fought in the bush wars that raged from South Africa up to Angola and ended in 1990, and the skirmishes which continued after the war ended. Like so many other soldiers, when their time in the military was done, they joined a private security firm that had an almost identical structure to the unit they left, including, it seemed, leadership. Over dinner, they could be heard speaking to their "command" on a satellite phone, in as formal a language as one would hear on any army base.

Bert greeted everyone on the team in his rich, Afrikaans brogue and give the group our safety and security briefing. "Some piracy reported last week, keep an eye out for unmarked watercraft. Some violence in the town last night. Word of two technicals spotted just outside Soyo."

Bert confirmed our work plans for the day had not changed: one team wet (drill rig out on the river), one team dry (drill rig and vehicles on land). The times and places were clearly defined so that, in an emergency, they could round everyone up fast. Our locations shifted often—some days both teams were mobile, chasing new sites—but the protocols stayed strict. On some projects, these preparations would seem like overkill, but not here. The war only officially ended in 2002, just three years before, and during the worst of the fighting, the camp security had been overrun. Armed men breached the base. People were shot. Women were attacked. Some workers vanished and never came back. That history lingered, thick in the heat, shaping everything we did.

After our meeting, one group would set out for a land-based drill rig. These guys drove a spinning bit deep below the surface to retrieve narrow columns of earth—core samples used to assess the ground's stability and load-bearing potential. This was crucial data for building a pipeline that would run from the LNG plant to massive storage tanks; an industrial artery stretching for hundreds of kilometers and wide enough to crawl through.

The crew was comprised of two Americans, Terry, a youngish guy from New Orleans with a southern drawl that snuck out between spits of chewing tobacco, and Lance his his older lead driller, a potbellied Texan who could tell the type of rock or soil just from the sound of the drill. I loved watching them work. They would pull the steel pipes off the rig and remove the soft cores from within. Each core was labelled for depth and location, then placed into long wooden boxes for closer investigation later, in the comfort and shelter of the lab. The men would reposition the steel, feeding it down the hole again as the machinery bored deep into the earth. The size of an ice cream truck at its base, the rig was controlled from a seat on the side, where the operator used GPS for pinpoint accuracy. Above, the drill boom rose like a sailboat mast. It was a tangle of cables and hydraulics, a point of curiosity for passing birds and, at dusk, bats.

The other team, lead by Tiny, whose sheer size belied his nickname, made their way to the drill rig anchored above the river. There, they pierced through layers of sediment to bring up soil for the geotechnical mapping of the future LNG plant. As with the land team, this involved sampling and analyzing rich clays and silts, providing engineers with the information needed to design stable foundations for the titanic structures to come.

Both projects had their risks. Our sea-based team had water dangers, piracy, and the constant menace of fast-moving storms and lightning; columns of rain torrenting down as if someone had thrown open a valve in the sky. The boys might see it coming soon enough to set the rig and jump on their boat back to the base. Sometimes they worked through it, swinging steel and bagging samples while the downpour fell in sheets. It was not for the physically weak. Once, Tiny, in a fit of late afternoon frustration, threw a wrench the size of a lacrosse stick one-handed across the deck. As a gesture of peace, I walked over to pick it up but almost couldn't lift it off the steel deck. He sat there, looking out across the water, letting the rage flow through and out of him like the sweat tumbling down his face.

One day, Tiny and Thomas Joseph, the only two Black members of our field team, left the water and came out to help the ground crew. Not far from us was the remains of an immense oil storage tank. It had been rocketed during the fighting and, half full of oil, the steel spent weeks burning to the ground. It was the size of a football field and now looked like a modern art installation, a strange metal wave atop a concrete plinth. We stood on the edge of the foundation, cold pop in hand, and gazed at the ruins go up in flames; wondered if people stopped to watch, or if they were too busy running, children and belongings on exhausted and terrified backs, trying to stay one step ahead of the war bearing down on them. Tiny could not decide how he felt about it all: the burned-out oil tanks, the modest houses riddled with bullet holes. "This could have been me, my children, living here in this place," he said, rubbing his shaven head with his huge hands.

Threats still lurked: hostile locals or militants, a landscape crawling with the snakes and scorpions, or malaria and other diseases and viruses. In March of 2005, the CDC identified an outbreak of Marburg in Uige province, just east of us. Marburg is a deadly and more transmissible version of Ebola, and because of the flow of people and workers into our area from nearby villages, we were on alert for infection.

The real danger for the ground team was the very thing that had plagued the locals for decades: landmines. Wander any village in the country and you see adults and children using makeshift crutches or wearing ingenious homemade prosthetics. Africa Watch estimated that Angola could have 1.5 to seven million landmines in its soil. Thirty-seven different kinds, everything from tiny anti-personnel mines to large anti-tank mines. The signs were everywhere: red and white skull and crossbones with "Perigo MINAS!" in bright white letters.

On this project, various companies had full-time mine removal units. Some were private consultants paid by the land cleared, others were part of larger NGOs who brought in experts from around the world to support and train locals in clearance and disposal. It was a project within a project, and we could see the teams working away in taped-off fields and roadsides all over the countryside. When their work was close to us, as it often was, we absorbed their staff and worked together to ensure our safety as a group. It was surreal to be taking notes on the hood of a truck and hear an explosion a few fields over. Arms waving that everyone was fine: they had found a damaged mine and could not remove it securely, so they gave it a "pop," a small charge meant to set it off or disable it.

The clearance team provided us with a map showing which areas had been cleared with tiny black dots. These appeared inside light red zones, which suggested active mines or unexploded ordnance. The sheer number of dots was breathtaking—thousands and thousands. The access road was marked with red cones and yellow tape so that you would not veer off the route deemed safe.

Some afternoons, I logged core with Terry and Lance. We lined up boxes, writing the depths and identifying hole location numbers on them. I would look up to see the clearance crews, face down in the grass, painstakingly lifting landmines out of the shallow soil, making their notes and mapping their own timeline. We always stayed far enough from each other, so that the tremors from our drilling would not put them at risk.

When the land-based drill rig was working close to, or beyond, the perimeter fence, we had extra security. It was out here that I met Samuel, a guard assigned to our rig. He was slight but tall, with an easy smile. I went over one afternoon, lunch in hand, to sit with him on some stones. Samuel was very shy about his appearance, always positioning himself so that his left side was turned toward you. The right side of his head seemed to bend at a strange angle. It was not until you got close that you could see a narrow gap running several inches from the top of his skull, a strip that light passed through.

I had seen enough injuries to not respond when we first met, but eventually, I asked him about it. He spoke matter-of-factly and without drama. He was from a village south of Soyo. During the war, men had come to his village, attacked and killed people, and then took him and other children. They were trained as soldiers and fought for one of the rebel groups. A few years before the war ended, he and some others were rescued. But as this rescue was happening, some rebels began butchering the young soldiers. Samuel was struck in the head with a machete. He mimics the angle with his hand, grinning, then shrugging. God wished him to be alive, he said, so he stayed alive.

They were settled in camps set aside for child soldiers and re-trained. Some became cooks, some labourers, and some security guards. I asked Samuel how old he was. He made a face of concentration. He thought he was twenty-three. "Okay," I said, "How long were you with the rebels, how long were you a fighter?" This he was surer of, "Eight years." They took him when he was about eleven. I asked about his family, regretting my question the moment it escaped my lips. He got quiet, a shadow passing over his face. *Todos mortos.*

I spent many days running tests in an air-conditioned geotechnical laboratory on the second level of a warehouse, a crumbing edifice that had not seen a drop of paint or repair since it was built. It was the last in a line of mostly empty warehouses spaced out at the far end of Kwanda Base, away from the busy port and the supply ships that were loaded and unloaded by cranes day and night.

The lab was a small. It held an oven for drying samples. The tables and counters were covered with equipment. Small machines hummed, pushing and pulling the soil in different directions, creating lines on a graph that spoke to the engineers, whispering the secrets of the land's history and potential. Among other things, we were looking for signs of soft or shifting ground that could undermine a build, or water tables that might complicate construction. It's a search for the invisible architecture below—what will hold, what will sink, and what must be avoided. Soil mechanics is a world all its own, and I spent hours extruding soil from Shelby tubes, the long steel forms brought up from the depths plumbed by the drill rigs. Extrusion is an art, and you must handle the samples as you would an infant. To drop one would render it useless for tests; days of work and tens of thousands of dollars blown.

The warehouse itself, with its scarred concrete floor, had seen many purposes over the years. There were rooms and areas where we did not go. I was told on my first day not to open the door on the ground floor, right below the lab. On the second day, I opened it. I found a tiled bathroom, and a swamp of human sewage more than a foot deep. What surprised me most was the lack of any smell. There were no flies—anything trapped inside had simply died. I shut the door. Years later, I wonder if that room is still there.

One evening, as I was walking through the housing development, I saw Tiny sitting on the front step of the house. His face was buried in his hands. His shoulders shook, and he was sobbing. Sitting beside him was Wesley, who helmed our drilling operation on the water. Wes gave me a look and just shook his head.

Tiny lived in Miami and had a few kids. One, a boy in his early twenties, was in a movie theatre when a guy sitting in front began talking loudly on his phone. After a few minutes, Tiny's son asked him to stop. The person stood up and shot him. The shooter then calmly walked out, without a word. Their son was now in surgery.

Tiny breathed heavily while Wes spoke. The other guys were inside, sitting quietly and drinking beer, giving him space. We found out a few days later that the bullet did not hit anything major; Tiny's son would make it. I thought of him, half a world away, and thought of the guards walking the fence—boys who had spent the majority of their lives at war, even as children—and wondered how the pieces of this puzzle fit together.

A month later, I left Angola. I walked out onto the tarmac of the Soyo airstrip, and stood next to two Americans from the oil companies. As we looked across the dusty horizon, a flash caught my eye. Almost instantly, a long, slender mushroom cloud rose, all blaze and brightness. The men gasped. The explosion looked large. Maybe ordnance was being disposed of, though there was too much fire in the explosion for that. One of the men noted that there had not been a sound. I thought, *too far away yet. You will hear it in three, two....* The sound wave hit us like thunder.

The plane took off, late as usual. Below us, the base, helicopters coming and going, cranes and warehouses; just beyond, dirt streets, the patchwork of shanty town dwellings, and dark smoke twisting up, staining the sky. Angola felt forsaken, like a house that had been wrecked and then abandoned without a thought to fixing it. As the dust cloud kicked up from our lift-off, I could see kids waving from plastic and plywood roofs.

Winter Steel
Rainy River, Ontario, 2017

The sheet of steel towered as the crane pounded it, centimetre by centimetre, into the frozen ground. A multi-fuel heater blasted air at the base of the sheet, while a smaller electric heater targeted the larger one. A heater to heat a heater. That's what fifty below zero looks like.

Twenty or so contractors were clustered into small groups at the base of two huge cranes reaching into the bright red sky. They were working in a wide-open area bordered by intersecting treelines. The workers bulged with heavy coats and pants, faces hidden under hoods and hard hats, behind masks and goggles. Steam rose off every man, hanging in tiny clouds in the still, cold air. The setting sun threw orange fire across the horizon, painting brilliant colours over earth and steel alike: dusk in Northern Ontario. One crane held the sheet in place and the second crane hammered it down like a nail. This is how you build a gold mine in winter.

I stood next to my truck, watching the two cranes work in unison. The steel sheet—or sheet pile—was some fifteen metres long and two metres wide. Workers guided its foot, or bottom, into position while one crane suspended it in air. The other crane, armed with a large hydraulic hammer, then drove the pile into the hard crest of the ridge as the crew watched the steel sink with each blow. Wooden braces helped keep the pile at a right angle and steady. Nearby, where piles had already been driven in, crews were welding them together along their length, ensuring a solid connection between the sheets. The sheet tops, standing ten metres high in places, appeared like a dark rampart against the flares of welding. The strange sculpture would be pummelled deeper into the ground and then cut off at surface level, leaving only a steel cap visible.

The men were installing an underground barrier, a continuous wall to slow the water that would start flowing beneath our feet when conditions became warmer. Movement of water had to be reduced or even stopped, so that the gold mine could have better control over what flowed under the tailings' impoundment, otherwise known as a Tailings Storage Facility, or TSF. Only a small part of the ore taken from the earth is the mineral or metal you are mining. The rest becomes waste, either from the initial blasting and excavation of rock, or from the processes involved in extracting it. The TSF is where a mine stores this material, often amounting to millions of cubic metres. It was critical that water be kept outside this zone. Mixing groundwater and tailings is dangerous because it can cause widespread environmental contamination by releasing toxic heavy metals, creating acid mine drainage, and posing risks to aquatic ecosystems and human health.

Although you could not see it through the snow, we were already standing on a vast raised berm, a sort of low dam also designed to hold off water. But the ground was permeable, had enough spaces and cracks for water to creep through. So here we were, building a fail-safe: a subterranean wall of industrial panels joined like stitches across a wound.

Before any mine is developed, years are spent studying what surrounds it. Water, flora and fauna, local weather, the shape and shift of the land: all of it is recorded and forms the foundation of a plan. The trick is to understand, beforehand, how the mine's construction and lifespan will affect the local environment and those who depend on it. This process, known as an Environmental Impact Assessment, or EIA, can end in a rejection of a mining licence, or a much-reduced proposal. An EIA plays a vital role in the design of a project and how it progresses. Everyone involved in decision-making refers to the document and should have a clear understanding of its findings. This is not always the case, but it's the ideal.

That ideal was largely what my team of engineers was here to enforce. My company had designed many of the earthworks—dams, berms, roads—and a dozen of us spent our days going from place to place, watching and monitoring the progress. This role is called Construction

Quality Assurance, or CQA, and it can mean different things on different projects. We were the eyes and ears, ensuring key details were not lost or overlooked. Contractors often have their ideas about how to get to an end goal. We could be flexible if the final product kept the quality and intent of the plan. Other times, we had to be tougher and say no, providing contractors with another way, or insisting it be done as agreed. We make hundreds of these decisions a week. They can be very minor, or they can be as large as the project itself.

Bright lights lit up the work area in patches. Men huddled over coffee cups and around the heaters. Welders in full facemasks, and heavy overalls had already started cutting and fusing steel sections, swearing behind their masks that fingers were already numb in their thick, leather gloves. As metal turned to liquid, I watched glowing droplets fall to the ground, steam shooting from holes in the snow.

I stood to one side of the flash guard, a set of frames that held up opaque plastic to prevent the welding flare from doing any damage to passersby's eyes should they look at it. Even peripheral viewing could harm you. I held up a pair of goggles every few minutes to watch the careful bead being laid across a seam. It seemed even cold fingers could make art of the task at hand.

A large parka topped by a hard hat approached me, the face obscured by the heavy fur liner that kept ears and eyes intact in this wind. Tim was the contractor leader, and he was doing his morning rounds. The child of Icelanders who had moved to Canada in the eighties, he and most of his team lived in a small community in Manitoba, where they learned how to work through serious winter. These contractors were doing all the major works on the site; shifting earth, constructing buildings and equipment, building the endless roads the site required. Even the large trucks hauling materials around the site belonged to them. They counted in the hundreds; everything that moved on-site was moved by them.

"She's a fucking frosty one. I heard it could get cold today," Tim yelled through his facemask.

"I heard that," I yelled back, "Until then, т-shirt weather."

His laugh carried on the wind and we both watched the hammer rise and drop, rise and drop, onto the head of a pile. The sound of the strike carried in the air like a gunshot. Progress in these conditions was tough; bits of metal could shatter. Driving frozen steel into frozen ground sometimes felt like a display of stupidity. We were Sisyphuses of the North.

We meet every morning, this crew of mine, young engineers, and older technicians. We checked on contractors and what materials they were bringing. We also wrote reports—lots of reports. We live and die by these reports, phrases, and photos bound together and fired off to offices around the world. We communicated what was happening on-site to people who were not here, and for posterity. It was, frankly, one of the most grinding parts of the job: spending each day watching machines and people erect structures seemingly out of nothing, then sitting for hours getting it all down. The demands of report writing have sent more young people running from this work than the loneliness of being away from home or the grueling conditions of camp.

The meetings that we took part in every day, with contractors, mining staff, and management teams, were another kind of hell. These sessions were ambushes, with deliberate attempts to trip us up and divert attention from mistakes being made. For contractors and some of the mining staff, it was all about hitting deadlines and cutting costs—a stark contrast to my team's laser focus on getting the work done right. As the team manager, I was considered the outright enemy. Here, someone could give me a thumbs-up right before a meeting, only to cut my legs off in front of twenty people minutes later.

The mine manager, the top job on site, was a very smart and very difficult Australian I will call Chris. He was the hired gun, recruited to get schedules and budgets on track, and push the players into driving the whole train in the same direction. Chris had a long-practiced tone of voice that did not require amplification to instill dread. He knew where the edges of ethics and law lay and walked that line with his toes well over the edge. Despite all of this, I liked him but was on guard every second I was in his presence.

Chris would go through the day's big-ticket items line by line.

"Earthworks?" he would ask, and a fifteen-minute conversation would commence.

"Water Management?" and the same again.

He then went around the room person by person so that items could be discussed, grievances aired, and questions asked. Why did this team dump that material in my area? How come our incoming supply of tanks has been slowed down? When will the government inspectors come to visit?

This meeting was the reason I was up at four in the morning, needing to see what the night shift had accomplished so we could report on quality issues and progress. They knew I had walked the ground already with my sleepy-eyed night shift, and that reduced the potential bending of reality.

As the call went from person to person, inevitably the stack of issues and complaints climbed higher; that we were holding up the work, that specifications were being too strongly adhered to, that my person in the field didn't know what they were doing. By the time it came to me, I had a list of things to respond to. I usually hit the ground hard and fast.

"Ok, but the whole lift failed, not just one test. So, you are tearing it out, today," I would say, responding to a layer of soil that had been placed in a rush, ignoring quality standards.

"Sure, but the material failed the gradation test, for a *third* time. Find better material," I would counter, in response to ignored lab test results for soil that was used anyway, against the rules.

"So, your dozer operator hit another instrument out on the dam crest. At this rate, I am going to buy stock in that instrumentation company, as we are breaking one out of every three. The seller must love you," as an answer to yet another expensive and important measurement device that was run down by a large bulldozer, a regular occurrence which set everything back.

"No, Brian, that was not the c-word he used. Your guy is back in the truck shop, or he is off-site, I don't want to 'c' his face again," in response to some words between a machine operator and one of our field technicians.

Mexican soap operas have been written with less material than these meetings. The contractors wanted to get paid for building a thing with as little interference as possible. It is so easy to let specifications slide at three in the morning in thirty below zero, but that is how dams fail. So, my brave soldiers stand there all night, reporting on the work as it happens. There are people for whom this part of the process is enraging, an unnecessary obstacle, but the failures of quality make for bigger failures later.

Then I was back out the door, into that crystalline morning and the parking lot of trucks steaming with engine heat and frosted widows.

I drove out to check on the pile installation and some drill rigs. It was the beginning of another long night shift; I would likely grab a few hours sleep in there somewhere but would be out here most of the night. My day-shift staff were all back at camp, either eating or in their rooms; my two night-shift people were out here in the cold, watching the twenty work fronts on the go.

When I set out from the office, the sun was still streaking the sky, and as I drove past the dense tree line just past our office complex, I heard something through the slightly open window. I stopped and parked. It took a minute to make it out, but the howling of wolves came across the air. It was a stark, lonely sound, making me feel even further from the world than I was. I drove on, wondering where they were going.

The crane was busy, the hammer slamming down. I continued to watch the piles, men working fast, steam rising from heaters and from the constant welding. Bright fountains of sparks reflected off the dark steel, creating shadows, ghosts of the night shift. The line of piles ran on for hundreds of metres, steel sheets sticking up about a metre. The new ones were much higher, standing ten or fifteen metres in the air as they were placed just deep enough to hold them upright while surveyors checked their angles and direction. The constant hammering shook the ground like a war drum.

I drove further out, cutting through the forest on a road laid across the frozen swamp. I saw the lights of the drill rig through the trees; a pillar of brightness, a straight narrow shaft of luminosity created by ice crystals in the air. I came upon them: men greasing drill steel and running heaters. The rig was set up to do geotechnical drilling for some upcoming mine expansion. The geologists needed to know what sort of rock was here, and if it contained the right mixtures of minerals they were hoping to find. These sorts of activities were a constant on these sites; the mine was always looking to expand its production, and if they were sitting on some copper or bauxite or, in this case, gold, they wanted to know, so decisions could be made if it was worth going after.

A worker was marking up core boxes, his parka and furry hood preventing me from seeing who it was. I could feel the cold seeping through the truck door. I parked and watched for a while, taking notes on the location and time. I stepped out, wearing thirty pounds of boots and clothes. I thought about the jungles I had sweated in, moments when all I wanted was a brisk cold to take the heat away. Well, here it was.

I spoke to the driller; he yelled back through his mask.

"Almost eighteen metres! We should be hitting that lower clay and silt soon, then we are out, next hole tomorrow!"

I nodded and took notes. Pen ink was getting gluey. I would have to finish in the warmth of the truck.

The rig stopped spinning and quieted down as the hoist lifted another piece of drill steel in the air, the helper adding just a bit more grease to the end to help it in the cold. The helper finished and turned to me. His eyes grew wide from behind his safety glasses. All the men facing me stopped and looked past me, over my shoulder.

I turned, and in the clearing between us and the tree line sixty metre away, I saw them.

The wolves trotted out of the trees, their breath hanging in clouds as they made a direct line for us. I counted six, dark greys and browns and one with a tinge of red. Neither the lights of our work area, nor the noise, had given them any pause. Their eyes held fast; this ground was theirs, and theirs alone. I have never seen wolves move like that. They did not stop or sniff, but crossed like they knew exactly where they were going.

One of the workers pointed at a lone deer, dashing through the drifts near the other treeline. It had slipped by us without being seen. The wolves saw it and were just taking a shortcut. One passed within ten metres of me. I did not breathe. It was not fear, though a healthy respect was present. It was awe. My chest filled with its beauty and sheer wildness. I knew that I could never be that close to a wolf again. Every man stood silent for seconds after we had watched them slip back into the forest.

The contractors continued pounding steel sheet piles into the ground for months, never stopping for cold or wind or night. My crews continued through the winter, preparing for the melting and freshets of spring, when our world would be under half a meter of water and mud. Our problems would change then, and so would our solutions, as we tried to balance what the mine wanted with what was truly best for the project. Like all ventures of this scale, people came and left. Eventually, it would be my turn.

The mine would continue to build and expand, under the eyes of those who did the work and those who checked it. Sometime in the summer, the mine would make their first gold bars. This gold would find its way all over the world, into jewellery and electronics, dental work and satellites. Somewhere in the darkest hours of the night, someone like me would be standing by a truck, watching the spinning lights of equipment moving earth and building dams.

Prayer Lamps for Lost Souls
Santa Cruz de la Sierra, Bolivia, 2012

The smell of the forest floor mixed with smoke hanging in the air. Wind made a flickering sound from dry leaves. A toucan flashed overhead in a sudden arc; its wings and beak seemed too large to swing through the branches so easily, but it cut an effortless line and disappeared back into the trees.

Fire season had arrived early in Bolivia, and it stalked our operation: a gold mine tucked deep in the jungle near the Brazilian border. Our camp was modest; a handful of small buildings clinging to a hilltop, once surrounded by lush green canopy, now overlooking a landscape faded to brittle brown. The sun should have been shining down, but the whole place was often choked in soot. Staff had taken to wearing masks and respirators, goggles and head covers. Camp food reeked; we reeked. Red eyes were rinsed a few times a day. People returning to the site would bring boxes of Visine for friends.

It was not my first trip here. We had a field team of engineers supervising various projects on the site. But the main event, the object of our attention, was a thirty metres high tailings dam being constructed out of dense, stubborn clay we had on hand. It was engineered to sit between two hills, forming a broad basin to hold decades' worth of mine tailings. This was how companies now managed their waste. Fifty years ago, they would have poured it into rivers or lakes, or just let it run over open ground, with little thought as to where the material would go or what impact it would have.

The dam was nearly finished, and it wouldn't be long before I was sent elsewhere. But no matter how often I watched it, the process never stopped feeling a bit magical. Dump trucks hauled carefully screened soil and gravel, unloading it in piles atop the dam's crest. Bulldozers spread the soft material, then flattened it, leaving a layer that the same dozers would then drive across, their heavy tracks tamping the reddish soil tight. A compactor followed, pressing it all down again. A tanker truck would spray water to get the soil wet enough to bind further. And so, in half-metre lifts, the dam rose.

Decades of research has gone into learning just how much water and compaction a particular type of soil can withstand. This limit is referred to as "maximum compaction." Field technicians tested the ground with nuclear densometres or "nukes"—lead-lined machines the size of a large shoe box with a tall metal handle jutting from the top. Nukes measured soil density and compared it to previous tests. The tool told a technician whether the material had been packed to the right standard. Achieving this wasn't as simple as repeatedly rolling over a lift. You needed the right gradation and the precise moisture content. Without these conditions, compaction wouldn't be homogenous, leading to weak spots.

These dams aren't meant to stop water entirely, as some might assume. They're designed to hold back the tailings while allowing water to seep through at a controlled rate, preventing pressure from building up and weakening the structure or its foundation. The tailings stay in place, but groundwater continues to flow in and around. The same principle is at work beneath buildings and parking lots: water moves through the soil, just as intended.

My days were spent in meetings and touring the grounds. With roughly five hundred staff and support personnel, the operation was relatively small—nothing like the copper mines whose open pits stretch wide as small cities. Some of those are visible from space.

Leading the team was a friendly engineer from Cochabamba called Marcelo, identifiable by his ten-gallon cowboy hard hat. I had seen these lids in catalogues but had never encountered someone with the balls to wear one on a mine site. Marcelo was the head of construction, so, as the tailings consultant, I spent a fair bit of time with him.

We drove around the site, following roads cut into the forest. We were on the lookout for what Marcelo called "bad decisions." One afternoon, driving up a slope,

Marcelo noted that there was some training happening—workers were being taught to use a chainsaw by a local instructor. We could hear the buzzing before we saw the men. There were four of them, three in full protective gear: helmets with face shields and glasses, work aprons, long leather gloves that covered their elbows, and work boots topped by shin guards. The fourth man, the instructor, was in a different state altogether. Marcelo and I watched with horror as he finished giving a talk and then proceeded to slice into a downed tree with a large Stihl. He did this in a T-shirt, shorts, and flip-flops. We got out of our trucks slowly, to avoid surprising them. Marcelo could barely contain his anger.

The man, covered with sawdust, pointed to the cut he had just made, explaining something to the students, the running chainsaw in his other hand. Marcelo waved his arms and asked him to put down the saw. The three trainees looked at their boss and could tell that the training was over for that day. They would be getting their training, but not from him. Marcelo escorted him to his truck; we would not be seeing him on this site again.

We continued our tour, which had now become a weekly inspection to look for safety issues missed by managers. We drove not only to busy work areas but also access roads and laydown areas, open spaces where equipment and materials were stored until it was needed. Mines are made up of huge areas, and even after years on site, you can stumble across places you've never seen before. This makes project management complex—and allows things to happen right under your nose.

We soon found three young men in bare feet and no shirts standing under the raised box of a parked dump truck. Marcelo approached them. We could see that they were draining hydraulic fluid from the main piston. The piston had likely been seizing, not raising and lowering properly, and they wanted to fix it. But with no fluid, there was nothing to create pressure in the cylinder. That meant the massive box could drop at any time, crushing the boys. Marcelo waved the three away, his face reddening with frustration. "It is your life we are trying to save! Do you understand? Your life!" he yelled in Spanish. I hung back, watching the faces, first bemused, then fearful. The contracting company was from a town outside of Santa Cruz, and I suspected a replacement would soon be found.

Marcelo called the company's supervisor, his voice tight as he asked him to come by. We both took a moment to let the anger subside. Marcelo took out a cigarette to light, and then, looking around at the bone-dry forest, thought better of it and put it away. He sighed, and his shoulders slumped. I patted him on the back and told him to cheer up. These frustrations were part of our jobs. "Hey, we caught them before they got hurt," I said. "At least they will remember this, even if they keep doing it. They just won't be doing it here."

"Yes, not here, never here," Marcelo said. Keeping people safe was a constant battle, and not one we always won. On projects where we invested time and effort, incidents were rare, the culture was strong, and workers took ownership of the process. But on other sites, with weak leadership and inexperienced crews, the problems started immediately—sometimes with accidents on the very first day. The promise you try to make to families and coworkers is that, at the end of a shift, everyone goes home. That promise was not always kept.

We waited in silence as the workers sat on the ground, their eyes fixed anywhere but on us. Eventually, their boss arrived. He seemed upset, though I had a feeling it was more about getting caught than about the fact they had risked their lives. He was a short, angry-looking man who put on the faux politeness of dealing with senior people but then turned and treated his men like garbage. I had seen it before. Abuse flowed downhill here, and the best I could do was write a detailed report recording what we had seen, hoping that limbs, or even lives, could be saved from it.

Weeks passed, and the fires crept closer. Work began shutting down in anticipation of an evacuation. Officially, we were there for quality control, but, as always, mission creep set in. Soon we were responsible not just for safety, but for advising on when and how to clear out. Most of my team, from Peru and Chile, were sent home, and I remained with the mine managers to see things through as long as we could. Some days the wind held, and we could smell smoke but had nearly clear skies. Other mornings, I woke to coughing, lungs raw, and stepped out into ashy fog.

One of the guys found an armadillo lying on the footpath, a tiny brown fellow who seemed stunned by events. The guys told me he was called a Gran Chaco, but I thought he must have been a baby, he was the size of an iPhone. They brought him into the office and fed him water through an eyedropper. He chippered up and sat back in our hands, looking at us with bright tiny eyes, and craned his neck like an infant curious about the world. He was either used to being held or, in his exhaustion, resigned to whatever lay ahead. We wondered if he had stumbled out of the burning forest. Each day, we saw animals cutting through the mine site, fleeing the scorched air and the crackling chaos behind them.

In the final days, small planes passed low overhead. I assumed they were spotters, tracking the fire's movement and relaying its position. More and more people had been leaving. Soon we would lose the kitchen staff, and that would likely result in a full shutdown. The call came in the afternoon. My office in Lima felt that we had given what we could as support. It was time to leave, while I still could.

The crew was used to working through fires, but this was as close as one had come during operations. Talking to the local guys, they had never seen such a long dry season. Walking across the courtyard between our housing and the camp offices, the leaves blew across the grounds in the hot wind. It looked like a fall day in Canada, except the grass crackled like paper underfoot. A single mistake, a single ignition, and we would not have to worry about the incoming fire, we would have our own inferno right here. The mood reminded me of the film *The Killing Fields*, where a population waited for catastrophe: the strange, quiet moments before anything happened, a low-frequency anxiety pulsing through everyone. In our case, when the bad news came, it was not in the shape of a murderous Khmer Rouge column moving up the road. It was that the access road would be cut off the next morning.

I had a ride out with an experienced driver and Ricardo. Head of the mine's rescue and response team, Ricardo was also a seasoned firefighter. He was calm, supremely confident, and had a terrible sense of humor.

We were getting squeezed by the fire's proximity. Based on air reports, it was moving through a lower gulley, causing a straight-line advance that could cut us off and surround the camp. The plan was now to leave in late afternoon. Night had a double effect, which could go our way or hurt our chances. The cooler air could calm the blaze, slowing its progress and giving us a better odds. However, the shift in temperatures from day to night could also cause winds that would encourage the fire, with the added danger of changing its direction, making it harder to anticipate. Ricardo was in contact with two spotters on his radio who were regularly calling the Bolivian military. This was not Colorado or California though, and manpower was limited. Also, the Bolivian forces had inferior night vision gear which made flying after dark risky. Ricardo gave me a two-hour countdown. He would make last checks and radio the fire surveyors for the latest conditions. When he got the all-clear, we would leave.

I loaded my stuff into the Toyota Land Cruiser at the front office, made two phone calls, one to my dad and one to my office, and then waited by the truck.

Ricardo came charging out of his office, right on time, chainsaw and go-bag in hand. He had fire retardant overalls and heavy fireman boots. On his head was a rescue helmet with a full-face shield in the upright position. He piled his own gear in the back of the SUV and handed me heavy work gloves.

"We may need to pull some trees or rubbish off the road as we go. Your gloves will burn through, so use these. Don't worry we will be moving fast," he said, smiling.

The driver had his gloves and was watching the trees above, the hot wind carrying ash now, the smell overpowering. We sped off and blitzed through the camp gates without stopping for security.

Ricardo was on the radio immediately, calling out the truck number and type and the passenger list. I saw his phone on the dash and saw no bars. We were outside the bounds of the satellite relay tower. The forest was quiet, the dirt road narrow and hilly. Ricardo explained that geography could create paths where wind would funnel through low sections and the conflagration could gain ground quickly. Also, winds could help flames jump across the treetops. Everything was dry and it was the perfect fuel system for the fire.

We drove on and the sun was beginning to set. Ricardo told me that the road ahead was open, but the heat was closing in and we were still two hours from the worst of it.

Coming over a small rise, we started down and drove straight into a bank of smoke. The truck was engulfed and the driver slowed down, trying to use the road surface to guide his way. Ricardo shut the air off, but the burning smell filled my nose—my mask was not doing the trick. My eyes stung as I looked out from the back seat into the darkening world outside.

We broke through the smoke and crested another rise, where a clearing opened up, and there it was: plumes rising out of the forest, all around us, for kilometres. I could see flames off in the distance, but also bursts of bright orange—spot fires. This was not the raging hellscape of a Hollywood movie, but a strange landscape of blistering collapse and erupting shapes amid the hills of what was green forest only two months ago.

Ricardo was on the radio again, calling in our position. "It is not too bad," he yelled into the back seat. "There is a truck about five kilometres ahead of us and they are getting through fine, just a couple of trouble spots. They had to clear some road which helps us, but we will likely have to do the same. Don't worry, this is going to be fun. I mean, who gets to do this?"

I had to admit, I was excited. I liked seeing the risks the way a professional like Ricardo would. I'd been in danger before. It came with the job and the kind of travel we did.

But the instant that cold blade of fear slips into your gut is when you realize the situation's no longer in your hands, that the moments ahead will decide not just your fate, but the fate of everyone around you.

What I fear most is losing control; realizing I'm just a passenger, with no real sway over what happens next. And now, in Bolivia, that's exactly what I was. The failing light had given way to a grey murk and now our headlights lit the way. We were crawling into the fiery heart of it now. I could see flames along the roadside, and bright embers in the trees. I realized that the fire had passed through here and we were driving through the aftermath. Ricardo was talking steadily to the driver, pointing out trees that could come down as we drove. There were bits of forest hanging over the road, burning as we drove under them. The heat was creeping in through the doors, and sweat was beading my greasy forehead.

Ricardo turned to me. "We have a downed tree up ahead, not too big. I am going to cut it, and you and I will drag it off the road. Put your gloves on, it is gonna be hot," he said.

Another fifty metres, I could see the tree. It had fallen straight across the road; its trunk was cracking and igniting. I opened my door; the hot air was sucked into the cab like a vacuum. I felt like ten hairdryers were being blown into my face.

"Whoa!" I yelled as I crossed my arm over my face. Ricardo had already gotten the chainsaw from the back and was now cutting one end of the tree. Sparks flew as he cut through a blackened section of the trunk. I stood off to the side waiting for instructions. Looking up, I could see sizzling glimmers flying in circles, tiny burning murmurations. I watched as these micro-storms flew through the canopy, dazzling and alive.

Ricardo had already cut the other end, and the driver and I grabbed the burning log and dragged it off the road. It was only about thirty centimetres in diameter and had the forest not been burning, it would have been a non-event. We dragged it aside and the driver walked over to me and began patting my arm.

"En fuego," he said.

I looked down to see my shirt sleeve was smoldering. I'll be damned. I felt nothing. I patted it down. Ricardo had gone a few metres ahead and chainsawed away another section of another tree that he had dragged off himself. Then we were back into the truck.

This was, I was told, one of the worst fire seasons eastern Bolivia had ever seen. But it wasn't just here. Russia's taiga was burning, sending ash into the stratosphere, though no one knew how much; the numbers were classified. Australia was entering what would become a relentless cycle of destruction, years before the 2019–2020 season that wiped out millions of hectares and animals. Since then, wildfires have only grown longer, hotter, deadlier. Year after year, the records fall. And still, we act as if it's all under control.

So here we were, driving through a forest on fire. I kept thinking about families trying to escape this: no truck, no support, maybe facing landslides or violence on top of it all. That scenario frightened me as much as the ride itself. We pushed through scorched stretches where smoke thickened, then cleared, then returned again. We stopped only once more to drag debris off the road and keep moving.

Before we made it out, we passed through a small valley, a deep ravine that was engulfed in flame. Again, I could feel the heat through the door, and, for the first time, I considered the fuel tank. The driver commented to Ricardo and pointed at the engine light; we were heating up fast. Ricardo yelled that we were almost there. The check point was less than two kilometres ahead.

I smelled burning now. Was the road itself on fire? Fulgent brush blew past us. I wondered how far we would get if we had to leave the car; we had fire blankets and survival gear, but it was so hot. Ricardo, sensing my fear gave me a wide smile and a big thumbs up.

We kept driving until the roadblock came into view, flanked by Bolivian police and fire crews. It was a staging area set up beside farmland where the forest had been stripped away. We stopped and were immediately ushered from the truck by a fireman. When I turned, I saw that the tires were aflame. The smell had been burning rubber. How were they still inflated? On closer inspection, it looked like we had picked up charred debris and it was sticking to the softened tread. The tires looked like they were going to burst. The driver handed me my other bag and we shared a hug. Ricardo and I walked past the checkpoint. He had a quick talk with the firefighters and spent a few minutes on the radio. I saw that I now had a signal, though I had no idea where I was. Ricardo caught up with me.

"Ok, Gilmer. We are far from the trainline so no trains tonight. We will get you a taxi and you will make the drive overnight to Santa Cruz. The road is crazy but safe. Don't worry, no fires along there," he said.

We hugged and I wandered over to a waiting taxi. The drive back to Santa Cruz was nightmarish, a slow-motion Mad Max affair over rutted dirt roads, with tractor-trailers and actual tractors all vying for position in either direction. As we rounded a corner and started up a slope, I glimpsed a far-off hillside. Its entire canopy was blooming red, and, even from that distance, I could see branches combusting and soaring into the night sky like prayer lamps for lost souls.

We came down the hill, and I saw two pick-ups on the side of the dirt road. My driver slowed and a group of men stepped out. They were firefighters, their yellow and red coats reflected in our headlights. As our station wagon continued through, the driver waved, wishing them luck. They said "Gracias," the fatigue evident in their voices.

They marched single file, the weight of axes, shovels, and chainsaws resting on their shoulders. One by one, their headlamps clicked on. The jungle swallowed them whole.

Lake of Diamonds
Lac de Gras, Northwest Territories, 2007

Descending into Diavik is like landing on a distant moon, the world sealed in a hard sheet of ice and snow stretching in every direction. The plane circled a few times and then slid into a landing pattern which brought it into contact with a rough strip that shook the outdated Canadian North Boeing 737, one of the few larger planes that can touch down on gravel.

The cold that ripped through the plane after the door was opened caused a wave of grumbling. The cabin speakers came on and the pilot welcomed us, casually noting that the temperature outside hovered around thirty below. This was why we had to fly with our boots and parkas in hand. I had seen men in fleece zip-ups and sneakers turned away at the boarding gate.

The second shock as you left the plane, after the freezing wind pulled your breath straight out of your chest, was catching sight of what looked like a 1970s Canadian school bus. It was our transport to the intake facility. Steamedover windows, tattered green seats, and rubber floor mat: it was all exactly as I remembered from childhood. I was not a tall man, so having my knees bashing into the seatback ahead of me was a novelty, and I felt for the bigger gents who I could hear complaining, and who must have been sitting sideways.

We crossed the length of the island, passing what looked like a natural hill, but was actually a massive stockpile of excavated material from the mine. The twenty-minute ride to the main camp ended at a huge blue and white steel building. We were directed through the front doors, into what looked like a food court in a large mall, and lined up behind a checkpoint, happy to be in the warmth. We each walked through a metal detector, and our bags were put through an x-ray machine. It was very much like airport screening, except we could all smell the food being cooked in the huge kitchen across the eating area. Most of us had been travelling for at least a full day, flying in from all over the country, so we were hungry.

After going through intake, you were handed a room assignment. Mine was in the building we were standing in—near-hotel quality accommodations, by camp standards. Others weren't so lucky. Their assignment meant a long walk down an endless hallway, then a brief step back into the cold to cross the road to the contractor's camp—a smaller, harsher setup more typical of mining camps across Canada. Tiny bunk rooms, narrow beds, heat cranked to sweat-inducing levels. I'd spent months there before, and I didn't take for granted the upgrade my construction management role now afforded me. After years of dragging my aching body through one brutal site after another, I figured I'd earned it.

I walked out onto the edge of the tailings impoundment. It lay on outcrops poking out from the ice-covered surface of a lake. Diavik Diamond Mine sits on an island in Lac de Gras, north of Great Slave Lake and Yellowknife, in a frozen expanse just 220 kilometres shy of the Arctic Circle—a technicality, really, since the cold here rivals the North Pole on any day. As far as I knew, the plant site building was the largest structure in the entire territory.

The region was a maze of water and stone, making any attempt at building infrastructure, be it a mine or human settlements, challenging and costly. The rock beneath us was among the oldest in the world and was the very reason the mine existed. It was kimberlite, a rare, dark igneous stone known as the primary host for diamonds. The gems lay hidden inside it, needing no chemical extraction—just crushing, rinsing with water and ferrosilicon, then spinning through a cyclone where diamonds and waste rock split apart by difference in density. To reach this treasure, the land itself had to be reshaped. From the air, the mine appears as two vast open pits. Known as A-154 and A-418, they were gouged out over years under the watch of an army of engineers. Their terraced walls, called benches, spiral downward in geometric precision, like an inverted ziggurat. Along these narrow ledges, giant Komatsu haul trucks inch up and down. At the bottom, far below the former lakebed, massive electric shovels claw out kimberlite and ancient rock.

Our duties covered the entire operation, monitoring several projects at once. It was our job to assure things were being done to design, and when changes were needed, that those changes were met with the intent of what was required. You were a good guy and a bad guy in the same hour, every day.

One afternoon, I drove out to the far side of the island. The sun was making its brief appearance, casting a yellow light that bent through the glazed air. In the truck with me was Justin, a young geotechnical engineer who was monitoring the groundwork. We had been friends for a few years, and I knew him to be a funny bastard who I could count on in a crunch.

"Fuck man, it's cold. When do we get a project in Jamaica?" Justin asked.

"After you do your time. Three months in some chill and the boy wants to work under palm trees already. Earn your keep bro," I joked.

The wind kicked the truck side to side. Off in the very far distance was a high mark, what looked like a white mountain against the pink horizon. It was Ekati, another diamond mine nearly thirty kilometres to the northwest. The landscape was so flat we could see the tall stockpile of ore material. The illusion of it was that it looked close enough to walk to.

Justin and I were checking on the dam. Kilometres of rock were stacked in a long pile that would eventually encircle an entire area. In the summer months, we excavated the trench that would be the foundation for the dam, opening up the permafrost so that our engineered rockfill could be plowed into the gap, filling the trench and letting the permafrost re-freeze. But it was winter, so the dam was nothing more than a long, elevated section of ground. But today we arrived in time to see the performance start. D10—a dozer the size of a bungalow—began by scraping snow from the rock to prevent any frozen layer from lingering on its surface. The trucks then came out of the distance, like lit-up spaceships, bouncing gently on tires higher than two of our trucks stacked. Their size made them appear slow, but it was, again, the illusion of distance. The first truck came down the dam, then slowed to make a wide turn, stopping in place. The box began to rise, and 250 tons of large, blasted rock poured in slow motion off the deck and down onto the ground. It was thunder. We felt the vibrations of it through our truck seats 300 metres away. A couple of hulking masses slipped free and rolled out across the snow, but most of the load landed just where it had been directed. The truck gave a quick blast of its horn and began to roll away.

Now it was the dozer's turn. We watched as its massive blade pushed the material into place, spreading it with surprising precision. Just a few passes were enough to shape the deposit exactly where it was needed. From our vantage point, we could see the quality of the spread—mostly large, angular fragments with little fine material—and how the dozer was trimming it down to meet the required one-and-a-half-metre height. This careful shaping ensured the rock could be compacted properly and remain stable, without settling or shifting when additional layers were added. This was how these dams were built, one stage at a time. The slow, relentless rhythm that changed the very face of the earth.

Just as the dozer was finished pushing and cutting the layer down to size, the white, red, and green lights of the next haul truck appeared. We could hear the operators calling to each other over the radio on the dashboard, sounding like fighter pilots over the crackling speaker. The dozer gave the truck directions for where he wanted the next rock load placed. From this distance, the Komatsu 830E haul trucks moved almost silently. On a haul road, they had a way of sneaking up on you—you learned quickly to keep your head on a swivel. Fully loaded, each one topped 380 metric tonnes, powered by diesel-electric hybrid engines built to handle deep cold. Their hulking geometry looked improbable in the snow-filtered air: yellow and black boxes on wheels, with tiny wipers brushing away flurries from the narrow cab tucked beneath the box's overhang.

The sun had dropped further in its low winter arch, and the bright pink sun dogs of the artificial light towers across the island shot in a straight line up into the coming night.

The north provided more than its share of dangers; the first of these being the cold. Improperly dressed, forty below zero will kill a man in about thirty minutes; he is likely past the point of rescue after fifteen. Hypothermia and severe frostbite are an excellent tag team for death. Wind ups those risks substantially, dropping the actual temperature well past what a thermometer may tell you and adding to the air conduction of body heat. This is why you can feel chilly on a warm Caribbean evening standing in wet clothes in a breeze.

One of the biggest hazards was whiteout—near-zero or total loss of visibility caused by wind-whipped snow, turning the landscape into a blinding void. In a place where 300-ton haul trucks roamed, that kind of blindness wasn't just disorienting, it was deadly. The mine dedicated days to training, much of it focused on staying safe around heavy equipment. From the cab of one of those towering trucks, the scale and danger became clear: a vehicle the size of a shipping container could be parked directly in front of you and still disappear from view. Operators spent weeks in simulators and trained closely with experienced drivers before being allowed to operate solo. Even then, accidents happened. Fatigue, distraction, or a moment of confusion could prove fatal when you were behind the controls of something that massive.

But the threat wasn't limited to giants of the fleet. Even a simple pick-up had to be ready for the worst. A hard and fast rule was to never drop below half a tank of fuel. If you were caught in a whiteout and had to pull aside for a few hours, or longer, how else were you going to stay warm? Our pick-ups were stocked with emergency supplies: food, blankets, shovels, fire extinguishers, med kits. As a manager, nothing was more stressful than knowing someone on your team was stranded in a storm and there was no way to reach them; the very conditions that trapped them made rescue impossible. Weather forecasts were monitored constantly, a fixture in our daily safety briefings, along with strict tracking of each person's movements. Radio check-ins were mandatory any time someone changed locations or stepped out of a vehicle. When possible, the buddy system was enforced—because in minus forty, even something as simple as a twisted ankle could turn deadly if no one knew where to find you.

For newcomers, the rules could feel overwhelming. But most adapted fast or didn't stick around. Every project attracted people who thought life on site sounded exciting, only to realize it demanded more grit than they had bargained for.

Wildlife encounters were a routine part of life in the North, and foxes were by far the boldest. One day, while driving around the site, I spotted one: a vivid red shape trailing Wayne, one of our lead surveyors. Wayne was out marking the route for the new dam, seemingly unfazed by his four-legged shadow. Foxes had a habit of tailing surveyors, drawn by the movement or maybe the company, but they carried a high risk of rabies and could turn aggressive without warning. Wayne had fended off more than a few over the years—his stories kept us laughing around the dinner table.

A day later, I was at the dam. The sky was clouded in, but we were not near whiteout yet. Beside me was Jay, my cross shift, the guy who filled my role when I was rotated home. I was watching another haul truck roll into view, appearing like a ghost out of the drifts, when Jay bumped my arm.

"You see it?" he said.

"What, fox?" I asked.

"Nah, goddamn, look at that beauty," he said.

Then I saw it. Huffing and puffing its way along the dam surface, tossing snow to either side as it shuffled. A wolverine in the wild was something I never fully adjusted to; they were part wonder and part nightmare. Provoked, they are like a mini-grizzly high on meth. It came straight for our truck, and I was glad for the 3,000 kilograms of steel, rubber, and glass between us. The wolverine cruised past, and with my window down just a touch, I could hear the rasp of his breathing. He gave us the slightest side eye, just a passing glare. His tracks, left in perfect suspension in the fresh flakes, were already disappearing.

One day, touring the island in the narrow band of sunlight we had in the winter months, we saw other lights. They belonged to long-haul truckers who had driven the ice road from Yellowknife, crossing over 400 kilometres of frozen lake and the occasional patch of ground. The ice road was the one shot the mines had at getting what they needed—fuel, food, gear—before the thaw claimed the route. Miss the window, and your only option was to fly it in, at insane cost and with all the headaches that came with it. Diavik's airstrip averaged two Hercules flights a day, its schedule in constant negotiation with other mines and remote communities that depended on the same fragile network. Coordinating the logistics of it all was like running a troop movement, with supply lines dictating not just the pace of work but how much ore could be blasted and hauled out. Maybe that's why so many who took these jobs were ex-military.

The lights grew brighter, and sharpened into shapes: long trailers, flatbeds stacked with wooden crates, rolls of liner, and fuel drums. One flatbed carried the blade from a dozer, and, on the next, the dozer's engine and cockpit. These huge machines were sectioned off and shipped around the world to buyers who rebuilt them on site. The process took weeks. Every day, when you drove by the yard, you would see more pieces of the puzzle fitting together, cranes lifting parts into place, and flashes from the welders who connected them.

The convoy rumbled up the ramp onto the island. Trucks blew their horns as they came ashore, undercarriages crusted with frozen sludge and gravel, windshields smeared with weeks of grime. We blasted our horn in response.

Headlights still stretched out for kilometres across the ice, and it was half an hour before the final vehicle pulled in. Then the radio crackled to life: "Last truck off the ice."

Diavik was one of three mines the ice road fed: Snap Lake and Ekati being the others. I expected that, as the sun fell, there were still trucks out there, pushing through the last stretch toward those distant outposts. Not everyone made it. One year a dozer fell through the ice, taking the operator to the clear lake bottom. Men I spoke to said the radio call had been awful, a man yelling, then weeping from the cab of his truck. As the dozer plunged below the surface, fragments of crust rose and then settled, water pouring out over the gap where the machine had been.

Drivers speak about the way the road shifts beneath them. Speeds are tightly controlled, and progress can be agonizingly slow—any faster, and the pressure might fracture the sheet below. Sometimes it feels like the road lifts to meet you, released from the load of the trucks just ahead. Then come the white-knuckle moments. Engines fail. Visibility drops until all you can see are two faint running lights in the distance. Wind slams into box trailers, threatening to tip them. All the while, there's that awful soundtrack: steel frames groaning under strain, ice creaking and popping beneath the wheels. Some men and women drive the road every season, returning like sailors to the sea; others travel it once, drawn by the adventure and wages, retreat to Yellowknife and quit on the spot.

The morning was clear. I had already knocked out three meetings and done a full sweep of the site. Projects were moving, and with two days left before heading home, I was winding down. But I had a treat lined up first. Sitting in my truck, sipping piping hot coffee, I watched a helicopter warming on the tarmac. The pilot was doing a run-through, getting everything checked and making sure nothing important had seized overnight. The Airbus AS350 B2 had come to life without so much as a whine, and the sound of its engines was confidence-building. The helo was a thing of beauty, the red and white livery glittering in the sun.

There was a knock on my window and a young woman jumped in. Aanya was a geologist with the mine. An Indian woman who had grown up in Brantford, outside of Hamilton. She was one of those people who pored over rock like a librarian pores over ancient texts. Her dedication and joy were contagious.

"Good morning!" she said, cheering my coffee mug with hers. "Can you believe we get to do this today?"

I smiled at her. Enthusiasm was a trait I held in very high regard when staffing a project. Mike, the pilot, hopped out of the cockpit.

"Ok, I think we are on," I said. I switched off the truck and got our bags.

We stood off the pad until Mike waved us in. The engine was running but the rotors were not engaged. He sat us down and got me set up with the five-point belt in the back seat. He then shut my door. Aanya sat up front, dazzled by the instrument array with its mix of digital and analog dials and switches. When he jumped into the cockpit, Mike signaled us to put on our headsets.

"Good morning flyers, welcome to Diavik Airways, there will be no meal on today's flight," he said. Mike ran through a very brief repeat of the safety walk-through we had done in the office that morning, namely that in the event of an "unscheduled touchdown," we were not to exit the aircraft until the rotors had come to a complete stop. The ice now was very thick around the island, so nothing would be considered a water landing if we did go down. We were in full-weather gear, but the helo was warm and loud.

Mike started up the rotors and announce to flight control that we were lifting off. Diavik tower told us we had a clean sky, with no incoming or outgoing aircraft, and no blasts planned. Blast operations occurred at intervals around the site, and activities would be suspended during these detonations. Using explosives to break apart the earth has been a staple of construction for centuries. Things got a lot safer after Alfred Nobel brought stable, reliable charges to market. But even now, there's no way to fully predict how ground will behave. Distance is still the best defense. Blasting, however, was new to a lot of pilots. I saw more than one get fired for trying to give clients a front-row view. Flying rock, shockwaves, and light aircraft are a dangerous mix.

We were up, snow and ice spraying as we pulled away. Mike didn't waste a second. He dropped us low and punched it forward. The world opened around us—wide stretches of frozen land, endless sky, and sunlight. It felt like a controlled freefall across the top of nowhere. Aanya laughed into her mic. We both felt it: pure joy.

Mike took us for a spin over the mine: the huge buildings, the main camp, the deep open pits with their frost-coated walls and benches reaching down into the earth. We could see haul trucks climbing out of the pits, inching along the roads that wound round the edges, steam rising off giant electric shovels, shadows where the sun had not yet reached. The wind buffered the helo, shaking us in our sharp turn.

"Ok," Mike announced, "Let's go see some drilling."

He straightened out and sped toward a point in the distance. The landscape was so featureless it was hard to judge our altitude. We were well above the tree line, but even then, I couldn't tell if I was looking at a scattering of boulders or the crest of a ridge. Then a small dot appeared below, trailing a smooth line behind it—a snowmobile cutting across the frozen surface. It gave me a sense of scale, if only for a moment.

Looking out over the hard white expanse, I was reminded of the Dene legend about the seasons. The world was locked in endless winter. One day, while hunting, the first people came across a bear with a sack around its neck. Asked what was in the sack, the bear said it was the abundance of summer. They could not convince the bear to hand over the sack so they went back to their village and planned a feast for the bear, thinking they could get the bear to sleep and then steal the sack. But when the bear appeared for the feast, he wore no sack. Frustrated, the first people followed the bear home. While the bear slept in his cave, they found the sack guarded by two other bears. A battle ensued, and although three hunters were killed, the fourth hunter, mortally wounded, was able to tear open the sack. Sunlight and warmth burst forth onto the world and covered the land, trees grew, and strange birds filled the sky. From that day on, summer came to the Dene every year.

The legend seemed to echo some ancestral recollection of the last ice age, carried forward in oral histories. Now, another shift in climate was underway. I thought of this flat, wind-swept plateau, sparkling with the hardest substance on earth, and how the ice road lasted for fewer and fewer weeks every year, and how soon, how all this could become a memory, passed down through generations.

We began our descent, and I could see the drilling camp—a drill rig on the ice surrounded by containers and two large tents. We circled once and then descended where orange paint marked our landing zone. Mike judged the wind direction from his instruments and the flags posted around the camp. We settled down without so much as a bump.

In the drill tent, the crew were boxing and logging rock core. The drill was spinning through the ice, grinding into the layers below. The noise required hearing protection. I noted the well-marked "kill switches," in case of emergency.

No fewer than three fire extinguishers were visible. The driller gave us a quick rundown of the borehole, its location and the depth they had reached, about thirty metres below the lakebed. They were chasing deposits of kimberlite that could yield more diamonds and extend the mine's operational life. I took notes on the progress of the drilling and any issues they had run up against—stuck casings and a blown hydraulic line. Both were the bane of working in extreme cold. I'd seen heavy-duty steel buckets, as big as refrigerators, snap like porcelain in fifty below.

Eventually, we returned to the helicopter. Mike had been watching the weather. Rough stuff was coming in, but we should be clear to get back. We lifted off in a burst of powder, and soon he was back to carving through the air. It was all speed, noise, and motion—an adrenaline shot straight to the chest. Our camp emerged as a dark spot from the sky, black roads and piles reflecting against incoming cloud cover. Halfway there, Mike gave us a taste of wild flying. He banked hard left and right, the horizon tilting and snapping back, the exhilaration of it all fully framed in our wide eyes.

It was snowing as we landed. I helped Mike tie down the helicopter and cover it for the night. The setting sun crushed the horizon into fragments of red and orange, bleeding onto white.

Two weeks after I left the site, I heard something had slunk under the kitchen in the contractor camp and had been chewing away at wires and hoses, causing havoc with the systems and the electricity. Unsure how to deal with the situation, they sent in Charlie, a big Dene chef who also happened to be a skilled hunter, trapper, and hockey goalie. Suited up in full goalie gear and tied to a rope, he dropped beneath the floor to see if he could reach the animal. Onlookers cheered as Charlie was pulled out, screaming. In his hands was the culprit, still thrashing, having already clawed clean through his padding: a raging wolverine.

The Price of Birdsong
Brokopondo, Suriname, 2009

The headlights cut the dark, but only just. Along the narrow road, moving things fluttered back into the trees the instant the beam hit them. The jungle was as dense as any I have ever encountered, leaves and trunks and vines that stopped you like brick wall. And how it sweltered! As a child growing up in Nigeria, I thought I had a good handle on what hot meant. I did not. Hot was five large bottles of water a day, three guzzled and two poured in measured amounts over your head when you just couldn't stand it any longer.

I had recently upgraded my ride from an aging Land Rover Defender that had seen too many jungle miles to a new Toyota Hilux pickup. The Hilux was as fine a machine as one could hope for out here. The only issue is that Suriname lies on the Caribbean coast of South America and has nearly no usable transport routes, so goods come from the Caribbean islands. That included trucks and that meant right-hand drive on a right-hand road system. This created its own hazards, but I quickly adjusted, giving myself greater leeway and safety on turns and edges. Though there was a comedic bent to my driving for the first few days.

I was out doing inspections, the only staff member from my company on site. It's common for projects to be managed from distant offices, sometimes by people who've never even set foot in the places where the work is happening. That kind of disconnect, combined with poor planning, disorganization, and a lack of on-the-ground understanding, often means one person—this time, me—ends up doing the job of six. We were working on a gold mine, building a series of small dams to hold back the tailings left behind by production. One dam stood out. It was largest of the lot, a bright red clay pyramid wedged between two forested ridges. It was made from a blend of residual soils, saprolite, and laterite—soft, tropical rock formations that had weathered unevenly into soil over time. Where the dams ended, the treeline started, an impossibly high canopy from which the thundering night creatures played their orchestra. Some of those notes could kill you.

The crew was stationed in the middle of the broad crest, maybe 100 metres wide and 600 metres long end-to-end. They were shoveling sand into a narrow trench that ran the length of the dam—a vertical drain, or "chimney filter," meant to guide water down through the center and out the bottom. From there, it flowed into a blanket drain that carried it off into the jungle. This setup wasn't for tailings. It was to deal with the heavy rains that hammered this place on a regular basis. With that much water coming down, you needed a way to keep it from building up inside the dam and blowing the whole thing apart. I parked my truck on the crest and stepped into the night. The sound of the jungle was everywhere. The men gathered in two circles: one watching the work as the rubber-tired backhoe dumped sand from the small pile into the trench. The other was gathered around a central figure, speaking with exaggerated arm swings and an animated face.

I walked toward the group and this central figure stopped his parade and came straight for me.

"Mista Jeremy, how is denight?" he said, bounding in my direction.

His hard hat sat atop what looked like at least ten centimetres of braids; his work overalls were unstrapped on one side, mimicking hip-hop stars of a previous decade; and his gold teeth flashed back the bright plants illuminating the crest. This was Cool Cat. Yes, it was his real name.

Cat filled me in on how things were going. They expected to finish this section of the filter by the end of the night shift, which would be a win for the day crew—they could jump straight back into building the dam without losing time. The handoff between shifts was critical, one setting up the other for success or failure.

Cool Cat offered me a cigarette and I asked him about his family. He grinned and made a joke about having too many girlfriends to keep track of. He bragged about being a legend in the nightclubs of Paramaribo, the capital of Suriname, and how the ladies "wanted a piece" of the best dancer in the place. We laughed and Cat did a little step right there on the clay crest. Then, he called his guys over to give them some instructions with me there. He began to speak in a staccato yet sing-song language, head bobbing, hands in motion. Every few words seemed a mash-up—part English slang, part Dutch verb, and something that, to my ear, resembled Yoruba. It was music.

That music wasn't random. It had deep roots. The workers were from a famous and much-studied group of tribes referred to as the Saamacan, or Maroons. There are few countries with as complex and varied a history as Suriname, this tiny slip of land on the northern shores of South America. It was bordered by Brazil, Guyana, and French Guyana, and cut off from the greater continent by as inhospitable a stretch of rainforest as one could find.

Once a long-held Dutch colony, Suriname was rich in the commodities that drove European expansion: sugar, coffee, rubber, cacao, indigo, and some of the world's hardest and most prized tropical hardwoods. The Dutch seized control in 1667, and soon after, the British agreed to trade the territory for a small island on the northeast coast of North America—New Amsterdam, later known as New York. Suriname became a late but cruel participant in the Atlantic slave trade. Before the Dutch, the British had ruled the colony for a brief and especially violent period, an era now remembered as one of the most sadistic in the history of Atlantic slavery. Early plantation efforts had targeted local Indigenous communities for forced labour, but the combination of deadly working conditions, rampant violence, and waves of European disease rapidly devastated those populations.

It was a priest, no surprise, named Bartolomé de las Casas who persuaded colonial powers that they should abandon the use of Amerindians, and instead focus on tapping the most strategic resource of the time: enslaved Africans. The Dutch established a direct trade route from West Africa to Curaçao, where human beings were bought and sold to supply Suriname and other colonies. Those who weren't purchased were simply abandoned on the island, a brutal origin story for much of Curaçao's present-day population. In Suriname, the slaves were distributed across plantations owned by wealthy Dutch families and trading companies.

After Britain abolished the slave trade in 1807, pressure mounted on the Dutch to follow suit. They finally ended the transatlantic trade in 1840, but that only stopped the import of new captives. Slavery itself remained intact. It would take several more decades before the enslaved were granted legal freedom—though "free" is a relative term, given the generations that had endured forced displacement, starvation, violence, family separation, and the systematic erasure of culture and history. And yet, amid all that horror, something endured—even grew. A new language began to take shape.

Think of slaves fleeing into the dark that I stand in now, running headlong into this deep blanket of canopy and away from the burning torches of a plantation. Think of Dutch policemen and bounty hunters pursuing them, a chase that would last years. Few legal rights were so consistently upheld as the perceived ownership of another human, especially a working one. Entire economies were built on the capture and return of escaped slaves. But then think of generations of people and how they cobbled together dialects of their shared languages, filling in gaps with the Dutch and the Portuguese they were surrounded with as many were also escapees from colonies and plantations in Brazil to the south. Think of the groups escaping into a blank spot on a map, a terrible green wilderness full of beauty and danger, the music of birds blending with growls and calls of things that would bite, sting, and eat you. I imagine night fires, songs that were sung to ward away loneliness, evil spirits, and whatever lay waiting in that dark. As years went by, different groups coalesced, often bound by their collective past, what plantation they had escaped from, which leaders they now followed. Their language merged, creating six distinct dialects of Saramaccan.

Cool Cat was interrupted by the headlights of the survey team. Two men stepped out of their truck and began setting up equipment—survey instruments mounted on tripods, capable of capturing highly accurate elevation data, angles, and the precise limits of placed materials.

The data fed into a real-time digital model of the dam as it rose, tracking volumes, placements, and progress with precision. This system not only confirmed that the dam was being built to spec but also created a reliable record of what was placed where. It allowed us to cross-check construction against contractor reports and engineering plans—to verify that everything was lining up as it should.

One of the surveyors approached us, fist bumping Cool Cat and then me.

"Eh Gilmer, how is de night?" he asked.

His name was Pav, and his accent was a rich combination of Indian and Caribbean. He greeted Cat with a couple of short words that were Dutch, and Cat answered back in Dutch. He pointed across the crest to a few things, conveyed the lift locations he wanted checked, and talked about timing for the rest of the night.

I had my own job to do, so I headed back to the truck and hauled down the Troxler from the bed. Bright yellow and deceptively heavy, the device was used to test soil compaction—the degree to which the dam material had been properly worked and packed. The goal was to ensure the material could withstand the long-term stress of the dam's weight without shifting or settling. It's a standard test, used on everything from the foundation of a high-rise to the subgrade of a parking lot. The Troxler—officially a nuclear densometer—remains one of the most reliable tools for the job. I'd use it here to take "shots" on the red clay beneath our feet, checking if the material was solid enough to hold.

The men always gathered when this test was being done. It was a real-time reflection of the quality of the job, and a hurdle as well. If the test was a failure, more compaction, or even removal of the section, would be needed. If the test result was good, they could continue to the next layer of material, building the dam ever higher.

The crew needed to stand back while I took shots, and so did I, as the machine used mild amounts of radioactivity. It was far less than, say, the amount an x-ray technician was exposed to, but precautions were always taken, the machines were constantly accounted for.

The men gathered, making sounds as the Troxler beeped, numbers flashing on the screen. I recorded the numbers and compared the results to the required standards. All good, sighs of relief and smiles, and back to work. The lift would not have to be redone.

I drove back into the night with many more stops on my list. I liked taking my notes someplace quiet where I could indulge, unbothered, in the air conditioning of the vehicle. I drove three more dams over and had just parked when I saw dim yellow lights that did not belong to another truck.

For a moment, I figured it was just a couple of workers with flashlights, maybe finishing up some fence posting. I eased the truck forward to where the road ended at the far edge of the dam. A narrow trail dropped out of the trees, crossed the road, then vanished again into the dark. That's when I realized: these weren't workers on foot. They were riding two three-wheelers. Two men sat on one. Another man, with bags and shovels, sat on the other.

They stopped dead right on the trail. All three disembarked immediately, one so fast his bike rolled a few feet. I could see they were armed, and their faces were a mix of surprise, anger, and fear. They stood in the beam of my headlights; one had slung an AK-47 from his back to his front but had not grasped it. The other two stood holding machetes. I'm unsure why I was not more rattled. My truck was facing the wrong direction and turning it around on this narrow dam crest could not be done quickly or easily. Backing away in the dark with steep drops on either side carried its risks.

They stood in the light. The sweat on their brows and arms gave a sheen to skin covered in the day's work: dragging rocks and gravel by hand, hands digging deeply into sand, spinning sieves under a running waterline, or pouring buckets of dirty water over fine, glittering sediment. They were looking for that flash, that spark that drove men mad.

Artisanal mining, illegal mining, informal mining—it goes by many names, and it has been done this way for thousands of years, long before huge equipment and drill and blast operations existed. It always seems to end the same way—with impoverished men, women, and even children, toiling under terrible conditions in tightly organized pockets of misery, hidden from view or, just as often, in plain sight of authorities.

In some places, artisanal mining is community-based. Despite the significant risks, it fosters a sense of cooperation among those who take part. In other places, it is nothing less than modern-day slavery, with armed militias in the Congo, Burma, Venezuela, and elsewhere kidnapping children from villages to staff operations. This issue goes beyond the blood diamonds that are often the focus in Western conversations—stones that led to the creation of the Kimberley Process. That initiative, designed to certify the origin of diamonds and exclude conflict stones from the global market, was voluntary, but many in the industry say it dramatically reshaped the trade. Nonetheless, the mining of diamonds and precious gems continues across dozens of countries, feeding underground economies that, in some cases, bankroll atrocities. In the Kivus region of eastern Congo, armed militias are reported to use forced labor—including children and prisoners—to dig by hand through fragile, mineral-rich soil. These workers endure some of the harshest conditions imaginable: beatings, sexual violence, starvation, and execution are constant threats. It's a level of exploitation that ranks among the most extreme examples of human suffering in the world today.

There are glimmers of hope. Some organizations, like Better Mining, are trying to guide artisanal miners toward more responsible, safer, and better-regulated practices. In other places, however, these small-scale operations have destabilized regions. In northern South Africa, some abandoned mines have been taken over by the Zama Zamas—

heavily armed, illegal outfits that extract metals and gemstones, and strip out copper wiring and equipment, selling whatever they can on the black market. These men spend weeks at a time underground, never surfacing, working with crude tools and unstable explosives that occasionally trigger intentional cave-ins. Zama Zamas have been known to bury fellow miners after a strike; deliberately collapsing tunnels, knowing they can return later to dig out the drift and claim the riches from dead hands.

The name, Zama Zama, comes from a Zulu word for "those who take a chance." They are generally backed by well-organized crime groups. There have even been rumors of the police running gangs. Nightly operations are interrupted by the gun battles between rival groups. In images of men digging with pickaxes, you can see automatic weapons lying close by, in case the call to arms comes over the radio. The men themselves are often immigrants from countries like Lesotho, Zimbabwe, and Mozambique; people risking the journey to South Africa in hope of a future or, at least, a means to make a living. Some are cast-offs from closed mines in South Africa, unable to find work elsewhere. Zama Zamas have brought local communities to a standstill, compounding the strain on areas already underserved by law enforcement. Private security firms hired for protection often find themselves outgunned, and, at times, directly targeted.

Then there are the buyers. Across the Democratic Republic of Congo, Chinese traders dealing in rare earth minerals are a common sight—at roadside markets with rusted scales, lines of men and boys wait to hand over sacks of coltan and other ores. And this isn't just a scattered operation; it extends well beyond the edges of artisanal mining. Xian Jiang mining runs illegal dredging operations in Congo, under the protection of a Congolese army which reaps substantial profits from the arrangement. Dredging rigs cobbled together from containers and barges can be seen all over the Basako territory of Northwestern DRC, pulling materials like niobium and wolfram from the riverbed. This not only robs the country of revenue and taxes but also suffocates the river with tons of sediment, creating unfishable waters for locals.

So here I stood, in the middle of the night, with three scared and armed men. In Suriname, they are called pork-knockers, a word taken from the same activity in neighboring Guyana. The name is said to come from the fact that the miners would eat pickled pigs or wild boar at the end of a workday. These fellows did not look like they were thinking of dinner, though I was encouraged that their guns had not yet been raised. I had no confidence in my ability to get through to them in Taki-Taki (Sranan Tongo, a Saramaccan dialect workers spoke). Instead, I tried some Dutch.

"Mannen, het is goed, wees kalm. Ik ga, geen problemen ja?" I said.

This was me telling them, "We had no problem, I will go, everyone relax." They looked at each other. They were sweating, exhausted in forty-eight-degree heat. They likely had been shoveling for hours.

"Hé, hoe zit het met water? Wil je wat water?" I said.

Offering water were words I had used dozens of times a day while driving around, supervising work. I was always making sure our guys were hydrated. The three men perked up but were still wary. I slowly went to my back door and pulled out a plastic bag with five bottles of water. I stepped forward and placed it on the ground near the man closest to me, noting with relief that his rifle was now slung back over his shoulder. They didn't smile, but the man who took the bag raised a hand and said something in Taki-Taki I didn't understand. I nodded and tapped my hand against my chest.

"Wees voorzichtig, erg heet, niet ziek worden," I said, "Be careful, don't get sick from the heat." We went our separate ways. No worse for wear. I had used the bulk of what Dutch I had learned, and it was exactly what I needed.

Two of the men looked young, the armed one closer to middle age. Probably the old hand, the experienced miner had been digging through small spoil piles all over this site for years—and, before that, elsewhere. These men knew hard work, and aside from a bit of stealing which happens everywhere, I had never heard a whisper of them causing any real trouble for the local people or the mine. They were likely feeding and clothing families only a few kilometres away. They mounted their rides and sped off, the one on the back finally giving a wave and a smile, looking as relieved as I was. I drove to the next dam, closer to the ongoing work and in sight of some lights, and stopped. I let out a long, deep breath.

By then, I was only sleeping a couple of hours in the morning and a few more at night. I was managing field inspections, conducting tests, overseeing lab work, running meetings, and doing all the reporting—around the clock. When I was eventually relieved, it seemed the message had finally been received as four people came to replace me.

The night before I left, the camp chef, a Lebanese fellow from Montreal, put on a feast. A small group of us from the project showed up and we ate and drank and played cards until after midnight. These men, like me, had been all over. We shared stories of Peru, Congo, Australia, the Canadian Arctic, Russia, Iraq. Someone would mention a bar in Chile or Botswana, and someone else would pipe up, "Ah, does that Kiwi bastard still own that place?" or "Who knew the best Indonesian food was in Utrecht? Crazy." I was struck again by what a strange breed we were—those who left home for the far reaches of the world, carving out lives in remote, unfamiliar places. I couldn't help wondering how much longer I could keep up the nomad life, and when, inevitably, the wheels might come off.

On the bus ride back to the coast, I watched a Chinese road crew preparing the ground for fresh asphalt, using machines that dumped and packed gravel with mechanical precision. But then came the jarring contrast: rail-thin workers in torn shorts and worn-out sandals, raking gravel under the searing sun, their skin scorched and blistered by the heat. They looked up as we passed by, blank eyes staring at some point in the distance. Standing over them were two men, fatter, better-dressed, with umbrellas for the sun and water bottles in their hands. And at either end of the construction, armed policemen. It was as strange a scene as I had seen since arriving, in an already-complicated place. I wondered if those men would make it home or be worked to death and buried in the jungle.

I had a few hours to kill in Paramaribo before my flight that night. I came across a bird shop. From within came a sound like a rush of joy: high, bright, layered calls. I stood in the doorway, letting the chorus wash over me. No symphony could match it, no orchestra was so intricate or alive. Inside, delicate creatures flitted between the slats of their wooden cages—black and white, blue and yellow, full of motion and music.

In Suriname, songbirds hold a place of honour. The most talented can fetch thousands of dollars, along with prestige and social standing. Their competitions—where judges score the complexity and frequency of each call—are broadcast nationwide, and the winners are celebrated like athletes. As I stood on the street, I watched the city flow past: Saramaccans, Indonesian students, Indian labourers, Dutch tourists. Conversations sparked around me, each one a mystery until the first word gave away the language. I thought about the layers of history here, of cultures lost and cultures fused, of identities both fractured and whole. And I thought about the quiet steadiness in the people themselves—a kind of grace that seemed to thrive even in a place that had never made life simple.

I looked back at the birds and thought how odd a hobby for a people with such a history: to keep, in little wooden cages, creatures of flight and song.

How to Fix a Mountain
Ancash, Peru, 2006

I WAS STANDING ON a platform, looking down at the tailings dam. It stretched for more than a kilometre, a vast wall of earth and rock filling the space between two mountains. As a member of the construction management team, I'd been coming to this copper mine for years, watching the dam take shape. Its structure pressed upward and outward with each visit, creeping higher along the mountain slopes to hold back the unrelenting surge of tailings that poured in through thick steel pipes tracing its edges.

But there was a problem. During the site exploration, we discovered that one side of the valley, higher up along the slope, was made up of karstic rock. This type of limestone is riddled with natural cavities and channels, raising the risk that water and tailings could seep through. To reduce that risk, we decided to lower the permeability of the slope the dam was built against. Put more simply, we needed to alter the rock so that less water could move through it. There were various ways to accomplish this, some more elegant than others.

The design team in Canada and Peru decided that we would use a method called grouting. This can mean many different things, but in this case, we were drilling a few hundred holes all along the future abutment of the dam where the karst was found. Then we would inject grout—a fine mixture of cement, water, and different additives—into the holes at different elevations. The grout would fill the fractures and quickly grow solid. As multiple holes were done, newer grout would encounter previously injected grout, creating a sort of grout wall that would dramatically reduce or even halt water flow.

There was a lot of theory behind the plan. We were using non-traditional methods to determine which grout mixes, applied at which pressures, would best fill the voids in the karstic rock. Years of research and testing had gone into this, much of it done in a staging lab in Vancouver. Normally, grout is blasted into the ground at high pressure to forcefully fill the gaps. But that kind of brute-force method carries risk: it can fracture the very rock and soil you're trying to stabilize. Our approach was more deliberate: using lower pressures, carefully monitoring how the ground responded, what we called its "feedback." The idea was simple: if you listen closely enough, the mountain talks back. By tuning into those variations, adjusting our methods in real time, we believed we could achieve better results with fewer materials, less disruption, and far greater precision.

A ramp on the mountainside allowed crews to raise and lower the platform, a massive, window-washer-style rig outfitted with a drill, pumps, and storage tanks. On deck, workers stacked bags of cement and mixed buckets of chemical additives—admixes that would alter the grout's thickness and flow—feeding them into a large mixer beside the pumps.

Using different grout mixes, monitoring how much was being pumped, and how much grout the section drank in was called "take." Sometimes, you started to grout and there was no take; the rock was tight and would not allow you to inject the mix. In other locations, the grout would flow and flow, and we would change the mix, making it thicker with every change, until the pressure started to build, and we knew we were not injecting grout into some endless antechamber underground. The crew managing the pumps and injectors were specialists, mostly Peruvian contractors, with a few Canadians and other nationalities. Our role was to guide the process; to interpret the ground's response and direct next steps. We ensured the method stayed on track and the results stayed within target. At the end of each day, we compiled our observations and data into reports, sending them to both the mine and our team leads back in Canada.

I have experienced a few direct examples of the Anthropocene, the idea that we have reshaped the environment so much we now control its course. But nothing made it feel more absolute than this exercise: we altered a mountain. Not just added to it, but changed what it does. We did so with less effort and materials than had been done before.

Being out at the front of something, the bleeding edge of what was understood, was as engaged as I have ever felt doing this work. Years later, I heard a line in *The Martian*, the film adaptation of Andy Weir's brilliant novel, that stuck with me. Stranded alone on Mars and facing impossible odds, Matt Damon's character turns to his video diary and delivers a simple, defiant plan: "In the face of overwhelming odds, I'm left with only one option. I am gonna have to science the shit out of this."

We were sciencing the shit out of it. The grouting program took years to complete, with crews of drillers, engineers, and technicians pushing through grueling shifts, day and night. It broke people. For every two who stayed, one left, undone by the altitude, the hours, the isolation. For those who embraced the science, endured the effort, the reward was unmatched. It wasn't glamorous. But standing on a platform bolted into the mountainside, watching headlamps flicker in the blackness below, and listening to the steady rhythm of men carrying on through the night, we had discovered something close to devotion.

The Strongest Winds in the World
Puerto San Julián, Patagonia, 2012

My morning drive from Puerto San Julián to the gold mine in Patagonia was the best ninety minutes of my day. Sunrises here carried a different sort of light, splashing vivid paint strokes across the picture in front of you. The sky itself was set like a portrait against the framed land, blues and purples on one side, dimming stars on the other.

I cracked the trunk window a few inches, letting in a rush of air that I imagined had traveled from Antarctica to cool my face. Descending a coulee, I looked across at the small river that had carved this valley over a hundred thousand years, exposing streaks of coloured rock and soil along its edges. Roadsides were lined by fences, for cattle or to demarcate property. Horses would appear, unbroken and fierce-eyed. I could never tell if they were unkept or wild. They stood along the fence line or ran at full gallop, chasing the strongest winds in the world. I passed gates with signs warning drivers that if the mine was not their destination, they were on the wrong road. I drove past a quarry where a front-end loader operator, at the end of his night shift, waited for the truck that would bring his replacement and take him back to camp. The quarry lay like an open wound—layers of gravel, sand, and clay drying in the sun. We scratched at that wound day and night, taking what we needed.

Before the mine gates came into view, there was the lake, a low blue-black body of water. As I topped the hill, I always saw them: hundreds of flamingos standing statue-still in the shallows, occasionally stooping over to pick at algae. Even after hundreds of passes, their presence shocked me—bright pink dots against the brown and yellow landscape. It was as if a street artist had, in a moment of inspirational frenzy, thought, *What is the weirdest thing I could put out here?*

The gold mine was a series of deep pits scattered across the Pampa, intersected by buildings, a camp, a truck shop, and a flat excavated area built into the side of a hill. Our team had numerous responsibilities, but my primary focus was the construction of a heap leach pad. Heap leaching is a process where crushed ore is laid out in layers over a large area. Chemicals, including cyanide, are released from overhead pipes and seep down through the ore, causing reactions that make gold leach out. This material then runs downhill and is collected and pumped out to the refinery for processing. Preparing the ground for heap leaching involves installing berms, troughs, and pipes. It creates a grid of cells, like a giant chessboard. A synthetic liner covers the pad area. Then, the ore is placed on the liner.

If a project has the space, heap leaching is an effective method for extracting valuable metals. Of course, it has drawbacks. One of those is the use of cyanide. Its versatility has made the chemical a part of many industrial processes, including the production of paper, plastic, and textiles. But anyone who has watched lots of spy films is familiar with the lethality of even small amounts of the toxic substance. It's no surprise the design and construction of these facilities is so gruellingly exact.

My team was made up of an international gathering of technicians and engineers from several offices around the world, all of whom had deep expertise in large-scale liner installation and collaborating in complex environments. The group in charge of the installation was a rough-and-tumble outfit from rural Argentina, who looked as likely to brawl as to build. But they were a talented bunch and eager to prove it. Everyone thinks they want to travel the world and do weird jobs in far-flung places. And indeed, there is a thrill in getting the call—a project in crisis, a narrow window to fix it. A plane ticket arrives, and you realize trusted colleagues have been summoned too. You cross mountains, rivers, deserts, and tundra, only to step into a room where they're all waiting. It feels cinematic—an *Ocean's Eleven* camaraderie and the satisfaction of be-ing one of the chosen. But when boots hit the ground, it is astonishing how many back out. I have worked with ambitious young engineers who fold up and flee, sometimes in tears, after a month away from home. This work is not for everyone.

In the early mornings, my squad would walk the ground, across thousands of square metres of prepared clay, liner installation, and ditches. Notebooks in hand, we would point out defects, note their locations, discuss causes and repairs, and observe production rates. We wandered in small groups; the alien landscape of the construction made small by the endless sky above, the winds reminding us we were just a waypoint to somewhere else.

The contractors would have several crews running simultaneously, laying down large rolls of protective liner and cutting it to be welded onto existing sections. Thousands of tests were run, patches cut, photos and samples taken. The radio called us constantly, sending us over to teams fusing liner sections together. Data was gathered nonstop, crew members hurrying back and forth to keep pace. Conversations were quick—numbers shouted, understood without explanation. When a number sounded off, heads snapped up and answers flew before anyone had time to ask the question. There is a satisfaction that comes from operating in an environment where competency reigns, where people know what they are doing.

When incompetence crept in, it brought grief in all forms, from the hassle of rebuilding a structure due to a simple oversight to the risk of losing lives. On the liner, a missed tear, skipped test, or botched installation could let leachate—cyanide included—slip past the protective layers and seep into the ground, maybe even the water. Preventing that was the point of everything. That, and not losing any gold.

Now and then, I glanced up from the liner to spot guanacos trotting along the mine roads. Also a common sight were Darwin's rheas, ostrich-like ground birds the local Patagonicas called Ñandús. Fearless and curious, they'd stroll right up to your truck as you climbed out, usually in the parking lot. They didn't seem drawn by the smell of food so much as the shelter our buildings offered from the wind. We'd look up from our desks to find one of these massive birds peeping in through the office window. The usual response was a chorus of "Ñandoooooooooo!" from everyone at their desks.

Unlike most projects, housing here was unusually fluid. For the first few months, with no accommodations available at the mine, we stayed in a hotel in Puerto San Julián. The camp had a few hundred rooms, but nearly all were occupied by on-site workers. While the locals saw the daily drive as a burden, we expats preferred the clear line it drew between work and rest—our own version of church and state. Senior staff were housed in spacious homes along the beach, built specifically for that purpose. Most had relocated with their families and spent five days a week, sometimes more, on site. As consultants, we followed a different rhythm; flying in for a few weeks, then out again, splitting our time between the mine, home, and other projects.

Most of the time, though, we stayed in town. Our hotel was modest but well-kept, a single-story, three-winged building just across the road from San Julián Bay. I would walk out past the front desk, and look across the waters of the inlet running in and out into the cold blue of the Argentine Sea. The setting sun cast a long shadow across the road to the beach, shaped like the Mirage jet fighter mounted in front of the hotel. Its sleek grey-and-green fuselage had been painted and repainted to withstand the harsh weather. Just below the cockpit glass were three stenciled silhouettes: British naval ships sunk during the Falklands War.

There's something uncanny about coming face-to-face with a country's inner narrative, the stories it embraces or avoids about its own past. In South Africa, it's the transition from Apartheid to the so-called rainbow nation. In Japan, it's the legacy of the Second World War, and how the country's role as aggressor is often glossed over in schools. For Argentines, it's the Malvinas, or the Falklands to the British—the disputed islands that still stir deep national grief. Walking around Buenos Aires, it's impossible to miss: graffiti and posters insisting the Malvinas belong to Argentina. It doesn't take long to understand that the war with Britain, the loss of territory seen as their own, and four decades of foreign control scarred the country.

But while taxi drivers and shopkeepers are all too happy to discuss the injustice of the outcome, especially now that sub-sea oil has been discovered off the islands, they become very quiet when asked about the time the military junta was kidnapping and disappearing thousands of young students and protestors in the dark of night, never to be seen again. History tells us that Malvinas was a carefully constructed distraction from the murderous actions of a dictatorial government. Yet decades after that government has fallen, the distraction remains firmly entrenched in the national narrative.

I had befriended the hotel bartender early on, a robustly built, silver-haired fellow named Gustavo. He could recite the names of the British ships sunk by French Exocet anti-ship missiles with ease, but his face darkened when I asked if he saw the war as a way to shift attention from the horrors unfolding at home. "Bien, it was a very complicated time, and for us, the war was not complicated, and you could talk about the war, you could rage against the British. You could talk to friends about the battles while watching the news at night, but you could not speak of the government, the generals. That would have been dangerous," he said.

I thought about how stories we tell ourselves change over time. As a child of a Northern Irish father, the sentimental anecdotes I heard as a little boy while walking the streets of Belfast with my parents were so different from what I saw with my own eyes as an adult. I think of being a schoolboy in Atlantic Canada, and the first, tentative steps of teachers trying to explain the history of residential schools in a jurisdiction still very much under the sway of the Church down the street. It is easy to criticize a country's failure to see itself clearly, but we all share that blind spot—both collectively and individually. This was how Argentina took hold of my heart and mind. Not just through its breathtaking landscapes, but through contradictions and complexities that drew me in with a magnetism I rarely felt anywhere else.

The windows and doors of most homes had steel shutters that rolled down to cover glass and doors. If you were wandering about and saw people sealing up their homes, you quickly learned to hurry back or find shelter yourself, before the weather turned and caught you in conditions that could become dangerous. It was on one of these outings that I met Benjamin.

I was running some errands, in case the weather stranded us in the hotel for a few days, when I saw a young man from my truck. He was setting up a small table on the empty sidewalk. He was wearing only a light windbreaker, and the wind was blowing straight through him. He was one of the few Africans I had seen in Patagonia since arriving eight months earlier. I parked and walked to deliver a letter at the post office. Walking back, I saw that he was still struggling with the table. I crossed the street and gave him a friendly "Hola!"

He looked up and returned a hesitant "Hola." I heard his accent immediately.

"Parle Français?" I asked. His face beamed. We exchanged names and then he burst forth with a swell of information. His name was Benjamin. He had immigrated from Senegal and was placed in Puerto San Julián as part of the terms of his stay. A host family had taken him in, and he was doing everything he could to find work. I was struck by how hard it must have been—an entirely new country, a culture that felt like another planet, and a language he barely understood. His demeanor was warm, his mood upbeat, but the strain showed in his eyes. His table was covered with trinkets: bracelets and necklaces, keychains, and bottle openers. I asked how business was, and he shook his head.

"The people have been so nice, but they do not know me. I am the stranger," he said.

The wind was now starting to blow, and I asked Benjamin if he knew a storm was coming. His face told me he did not. He established where his host family were, and I agreed to help him pack up and get home. As we drove up the main street, people were shutting their homes, leaves and paper were already blowing up the street in tiny twisters. Benjamin held his hands in his lap. I thought of friends at home who were nervous about their kids going off to school, and how this young man had crossed oceans to come to this place where he knew no one.

We arrived at the house, and I could see the family in the window. The father came out, and we brought his things inside.

"We did not know where he had gone. We tried to tell him of the storm, but he didn't seem to understand," Ricardo, the father said. I could feel the warmth of the home. I watched the wife lay her hands on Benjamin, gently, the way a mother would. For the first time, I saw just how young he was. The family invited me to stay for dinner, but I declined. I left my phone number, feeling far more at ease now that I'd seen where Benjamin was living. I gave him a proper hug and told him in French, "Call if you need anything."

Benjamin smiled and nodded his head. The children gathered at the door and giggled watching me switch from Spanish to French and back. They seemed madly in love with Benjamin, fighting over who got to stand next to him. I returned to my truck, passing a couple more people on the street. Both asked if I was heading home, concerned I might get caught in the approaching storm. I told them the name of my hotel and thanked them. Then I drove back and parked out front. As I walked into the the bar, I could see the Mirage jet on its stand, bouncing in a tug-of war with the gale. I could see the sky darkening with the storm clouds, off in the direction of the sea, of those islands with two names.

A week later, I was out on the pad, watching the contractors install liner in squalls that would soon be too strong to work in. I heard a call over the radio and saw some guys gathered in the lower corner of the pad. I drove over.

In the catchment ditch, at the base of the liner slope, was a baby guanaco, kneeling and mewing for its mother. The mother was about ten metres off to the side, unable to walk on the liner. Hoofed animals like guanacos, which are even-toed ungulates, have difficulty walking on smooth synthetic surfaces. Something about the contact doesn't allow them to have a good grip, so they slip and slide and can become injured as a result. I could not, for the life of me, figure out how the little one had wandered that far out, having seen their inability to step onto the plastic. Perhaps the little guy hadn't gotten the memo, but he was stuck and not looking happy.

Our resident biologist was gone for the weekend, and so I was faced with a team of technicians, engineers, and operators, all looking completely distraught. Honestly, I would say it did my heart good to see these big, tough guys beside themselves with concern. One brave chap tried to climb into the trench, but as soon as he got close, the mewing became a frightened howling. It was a bit of a fix.

I was becoming worried that the baby was going to exhaust itself past the point of recovery, or that the mother would abandon it. A familiar voice appeared at my side. It was Philemon, a Kenyan who joined my crew by way of Houston.

"Ah! The baby is caught. What a shame! Let's pop him out of there," he said.

I explained that we had tried to get close, but the panic that followed made us fear the baby guanaco might hurt itself. Philemon asked everyone to step back. We retreated twenty metres. I could see the mother guanaco paying close attention now. Philemon began cooing to the baby; a low lullaby of sounds that caught the young one's attention but did not send a jolt of terror through it. Philemon stepped carefully into the trench, which was only a metre deep with soft slopes and moved closer. The guanaco was mewing but not loudly; Philemon's voice seemed to have slowed everything down. Perhaps one man approaching quietly was better than a group of us all standing around noisily. The baby kicked out once or twice, its toes slipping on the plastic, and then it bowed its head down, perhaps in hopes of hiding. Philemon slid over and wiggled his arm under the baby's long neck, his other hand reaching under the legs to support it.

"Come little one, let's go see Mama," he whispered.

One of the guys was taking photos as Philemon carefully stepped out of the trench. The baby was lying docile in his arms, its head across his shoulder. The mother had stepped closer and was now waiting patiently along the edge of the liner. I had expected some sort of aggression, or perhaps for her to bolt, but she stood quietly, waiting for the handoff.

Philemon stroked the baby's neck as he approached the mother. He stepped to the edge of the ground and set the baby down. The tiny guanaco knelt, depleted, and the mother came to it, sniffing and licking. A sound came up from the nearby men. It was the sound children might make when presented with a puppy. Philemon was still grinning when he walked back to us.

"Please, we must now leave them alone. She will take her baby when she is ready," he said.

The baby let the mother nuzzle it. Eventually, it reached up to her. Fifteen minutes later, it struggled to its feet, unsteady at first, but soon finding its balance. Mother and child then trotted off, as if nothing had happened.

I thought of the circumstances: a group of locals and foreigners witnessing a near tragedy, and here, in the tail end of Argentina, a Kenyan stepping in and saving the day. Philemon insisted it was a non-event; his farm always had animals in distress. But for me, it was an extension of the relentless hustle that is project work. Such moments are what made it a privilege to do this—to being here, with these men and women, out on the edge of the world.

The weather kicked up again, sand and grit going airborne. In the distance, dust devils were forming. The guys secured their gear, and started coming in. I watched low clouds moving fast, riding on the winds that came from Antarctica across the burning sky.

Maps of Old
Katanga Province, Democratic Republic of Congo, 2010

The checkpoint looked abandoned. Clouds were gathering. Not the white ones that cap morning skies here in Katanga Province, but grey and black, rumbling across the sky like God's own cavalry. Afternoons in the Democratic Republic of Congo were muggy. Violent storms were the norm, lightning breaking the sky up into sudden patterns of shattered glass.

I kept watching the empty gate, and then saw the guards through the window of their tin house. Two faces waving at me from behind the steamed-up glass. Their eyes were large and panicked, their hands waving and pointing. One of the guards was making a funny movement with a finger—curling it back and forth as a child might. I had seen that sign before. It meant snake.

I thought about what it would take to cause these two, who deal with snakes all day, to be crunched up inside that small gatehouse. It was the equivalent of standing on a table when a mouse appears. My eyes followed their pointing, and I could see a shape on the ground through the dirty glass of my Hilux window, where raindrops had started to fall: a puff adder.

A puff adder is like a landmine. It waits in the grass, patient and lethal. Its colouration makes it hard to spot, and its venom is cytotoxic, meaning it attacks blood and tissues. Minutes after the bite, a victim will experience intense pain; bleeding gums and nose bleeds can occur shortly after as the venom goes to work. If the bite goes untreated, the laundry list of trauma that follows ends in a brutal and bloody death, often within twenty-four hours. Even if treated, loss of limbs, tissue necrosis, and other permanent health impacts are common. These snakes are nasty business.

Our fellow was coiling from the rain but seemed entranced by my truck, glaring up, nearly out of sight. The guys in the gatehouse were now grinning, holding up the clipboard. No, I was not getting out to sign in; they could take my plate number after we passed. I radioed security that we had a puffer at the third gate. No one should approach on foot. The project had biologists and animal experts to deal with such situations: people who could identify a butterfly flapping past or the call of a migrating bird on its journey from Finland. I recognized when I was out of my depth.

My passenger was craning to get a better look, a skinny Englishman we called Bacon. He was a PhD candidate who had a thirst for adventure and found himself packed off to the DRC. He was nearly in my lap when I pushed him back.

"We aren't parking, Bacon," I joked. I rolled down the window, just a hair, and saw that flat, V-shaped head staring right at me. You are someone else's problem today, cowboy.

Our main purpose in the DRC was to oversee the construction of a massive tailings dam. I led a team of engineers and scientists, mostly based in the UK, alongside a Congolese lab crew handling on-site testing and analysis. The dam, already in partial use, had been under construction for several years and had reached a height of about twenty-five metres. A dam like this begins with a small starter section, built to hold the initial flow of tailings. As the tailings accumulate behind it, the dam is gradually expanded—widened and raised in height to keep pace with the growing volume. Each phase of this expansion is known as a stage, and these are designed, planned, and constructed over the course of years, sometimes decades, as the mine continues to extract material—be it copper, iron ore, or gold—and generate more waste. The dam, in effect, grows with the mine.

The project stretched across several kilometres of green hills, with workshops, warehouses, trucks shops, and industrial sheds linked by bright red dirt roads—the only kind you'd find in this country. Piles of steel, industrial equipment, pumps and electrical boxes, and transformers all shared space inside fenced-in laydown areas to try and limit expensive gear walking off into the sunrise. Welders balanced on scaffolding, fusing steel beams destined to become massive new facilities, showers of sparks cascading to the sunbaked ground. Excavations showed the footprint of future structures. The mine was expanding, readying to begin producing gold as well as copper, and that web of operations would need to be housed.

Katanga was surprisingly calm and uneventful, given the backdrop of a country with one of the most violent histories in the world. In this region, peace often felt temporary, hinging on the changing winds and the temperament of whichever militia leader had turned his attention to a town or valley. I had worked adjacent to the DRC on many occasions, and like maps of old, with their blank edges and pictures of sea monsters, Congo was a place to be avoided. "Oh no, we don't go there," was the common refrain when companies were asked if they could staff a project or carry out a drilling program. The truth was that people around the world depended directly and absolutely on the DRC and the treasures in her forests and mountains. In every major conflict and period of expansion in modern history, the Congo has supplied something essential. Be it rubber for car tires, uranium during the arms race, or the coltan and cobalt of today's cell phones and satellites. One might be tempted to say, *So the country should be wealthy, right?* But I think we've long since let go of that illusion.

We drove in through the access road and stopped at the main office, where I dropped off Bacon, who had reports to write. I continued through the main industrial complex, past buildings so vast a city block would vanish inside. These buildings were basically roofs, supported by steel. Under each roof, drums spun in place. And inside these drums, crushed ore was further pulverized with heavy, softball-sized steel balls. The ore was smashed down into a kind of grist that would be the right size for the processes further on, when the copper and gold could be extracted from the rock with chemicals. The air carried a rich, sour tang from the activity. Stay long enough, and the scent faded into the background. But on returning, it struck you all over again—sharp and sickly, like the lingering taste of a passing flu.

I drove into the open ground around the mine. Not far away, I could see the mast of a drill rig, working just past the long roofline of a new part of the plant. I parked and approached, stopping at the safety tape to make eye contact with the driller and get approval to enter his area. His bright smile and thumbs up let me in. I made sure my hard hat and glasses were in place and walked over. I could feel the drill's vibration in my boots.

The crew consisted of three men. The lead driller was a Congolese man in his thirties. I will call him François, and he had experience drilling all over central Africa. People sometimes have an image of westerners pouring into a place and building these projects, but the truth is more complex. The mines were staffed by people from around the world, including from the country you were standing in. Francois was tall, the sleeves of his overalls cut away to reveal large arms that bore the scars of his work. His hands were worn, knuckles roughened, fingers scarred from being caught far too often between metal and gravity.

Francois watched and listened as the drill pipe drove its bit dozens of metres under the surface. One helper stood to the side. The goal was to better understand the subsurface—what kind of soil lay beneath us. We called it geotechnical drilling. It wasn't deep work, but it brought up core samples that could be tested and studied, offering clues about the ground we were building on. The designers needed to know how much stress the soil below could carry, or if softer soil needed to be excavated out and replaced with stronger stuff. As they worked, a third man ensured the soil cores were carefully laid out and labelled in the few boxes they had.

The core boxes: those were my immediate crisis. In the DRC, sourcing equipment was always a challenge, but these wooden or plastic containers proved unexpectedly elusive. And yet their importance couldn't be overstated. In some cases, projects and companies were required to store and protect core samples for years, both for further inspection and for legal reasons. In all my years on the job, I'd never had trouble getting hold of these basic boxes. They were usually just standard kit, stacked in some corner of the yard. But here, they were treated like gold.

Several hundred of these boxes had vanished from an off-site warehouse. A replacement order was placed, with new boxes set to arrive by truck, making the long overland journey from Zambia. But as with many things, the words, "Yes, the truck has left, should be there in seven days," are sometimes the last words you hear after the bill has been paid. In this case, the provider had a good track record, and the real possibility was that the truck had been stopped at the border, papers being shuffled from table to table for days, while the driver waited, cooked, and slept next to his vehicle, the final bribe having not yet been decided on. The driver may even be one of hundreds stalled at the border of a nation desperate for the cargo they carried.

This was the reality of managing projects here. Frustration served no purpose. For many Congolese, even simple necessities like mail or medicine came with an extra cost, discreetly expected by those in a position to withhold them, The other reality was the quality of the route. I had made the same journey to the airport in Lubumbashi: eight to ten hours navigating the slick, thick puddles of the main road. In some stretches, it was little more than two worn ruts down the center, shared with a chaotic mix of bicycles, cars, trucks, and buses, all vying to avoid getting stuck in the mud.

So, core boxes. Here in the green hills of southern Congo, I was meant to make them appear out of thin air. Francois looked at me. "You have a plan?" he asked. I did have a plan. It was not only a shoddy plan, but it also required daylight burglary.

In my travels around the site, I saw piles of different

equipment and materials. Mines eat up hundreds of millions of dollars of stuff when they're being built. Warehouse and laydown areas are full of everything from bricks to buckets, huge wooden rolls of cables and wires, giant electrical transformers, office doors, and miles of steel girders. One day, I saw a stack of plastic green roofing material, the kind that was made in that wavy shape to channel water off the roof, likely purchased for temporary buildings that had yet to be built. I was confident that I could line up the core in sections of this plastic roofing and then lay another sheet on top with bricks in between to protect the core. But I needed an elevated base to keep the sheets off the ground and out of the wet. Our storage area was nothing more than patch of dirt with a tarp roof. Rain would come, destroying the core.

The solution came when I saw a man throw broken wooden pallets into a pile. There had to be hundreds of pallets on this site, though they were also in high demand. The next three weeks were spent getting to office early, dodging security guards and warehouse managers, stealing a few pallets at a time and stashing them in my out-of-the-way drilling area. Over a decade later, my eye will still catch sight of a stack in a corner of a parking lot or against an alley wall, my brain still registering the contraband as a small victory to be seized. And the core fit perfectly.

Later, I started driving back to the camp. About fifteen minutes out of the gates, a few other trucks fell in line behind me, all from the mine and all headed back to camp or down the road. A mixture of dust and village fires that had begun to burn for the night. The orange shimmer created a shelf of light below the clouds which were now taking on the colour of the setting sun.

Staring at this horizon, I saw a motocross bike cut around me. The driver had the handlebars in a death grip as his rear tire spun and the bike nearly flipped from under him. His passenger, a bald man who looked right at me as they passed, had sandal-shod feet sticking out to both sides as if trying to balance their way out of the near crash. He was leaned so far back, that his ability to stay seated seemed an illusion. As they pulled out in front of me, I saw that the driver was wearing a white plastic bucket on his head, something a restaurant would order margarine in, and rubber boots. The bike picked up speed and when I looked at my own speedometer, I saw that they must have been going more than eighty kilometres an hour, on an uneven dirt road covered in potholes, ruts, and hoppers. They sped up, both leaning forward now, fully committed. They disappeared around a bend.

"This will end badly," I thought to myself.

Two kilometres ahead, the bike had spilled and both men were on the ground. I put on my emergency light and pulled across the road. The truck behind me did the same, and the third truck slowly pulled ahead of the whole melee and parked across the road in the other direction. All three trucks emptied, and we approached the scene. The bike lay in front of my truck, the front end crumpled as if it had run into a wall. The driver was lying facedown with his butt tilted into the air, looking like a large toddler mid-nap. His arm was broken at a terrible angle and as I approached, he began to make a sound between a groan and a gurgle.

The passenger was further ahead. He was sitting up, watching us, smiling, his eyes glazed over. He held up a hand, trying to wave me over, his wrist swinging like a pinwheel. He was trying to talk but seemed to have bitten his tongue and blood dripped from his mouth. I could see his sandals had come off, and his right ankle was smashed, as if it had been caught in a factory machine.

I dropped my first aid kit, grabbing gloves, work glasses, and a face mask. Ebola and AIDS were serious concerns here, and there was blood everywhere.

My guy, the passenger, was calm, the pain had not yet hit him. He was in shock and, I realized, was giving off the vapour of whatever moonshine they had been drinking. The smell captured me, momentarily, taking me back to basement poteen in county Armagh.

"Salut, quel est ton nom?" I asked.

"Ok, ok," he smiled back. I could not tell if it was shock or my terrible accent.

"Vous avez eu un accident, d'accord?" I told him.

"Oui, oui," he nodded. Ok good, he understood.

I gave him a very brief body scan, and I could not find any sources of blood or points of damage except for his destroyed wrist and ankle. Other than his bloody mouth, which I saw was a lacerated tongue, his head seemed utterly undamaged, a miracle of the first order.

He was breathing, speaking, smiling. I had heard stories of drunk drivers walking away from terrible accidents because the alcohol prevented their normal reflexes, so that they were supple, rubber dolls when the crash came. I looked back and counted no less than twelve metres from where the bike lay. The lack of pavement and soft road surface probably saved his life. One of the other truck drivers walked over. He said a few short words to the man in Swahili, and then to me in near-perfect English. "He's lucky, a bit fucked up, but he's good. Should be dead."

I nodded.

"Yeah, Jesus, I don't think he even hit his head," I said.

"Yes, looks good, drunk though. The other guy is no good, no good. He is broken up pretty bad."

I looked back and saw very different looks on the men gathered around the driver, who had been turned over onto his back. His chest rose and fell, and it was staggered, as if his body couldn't agree on how to breathe; his chest looked lopsided. A South African engineer was speaking to the man in Swahili, quite loudly, so I assumed they were not getting the responses they wanted. My guy looked over to his friend, and his bloody lip sank, causing more blood to flow onto his polo shirt. He tried to point with his broken hand.

"Ah, très mauvais," he said. His speech was a little more slurred now. I didn't think it was the alcohol this time. I checked his head again for an injury, especially one that was not swelling up, but again found nothing. Pain began to cross his face, and I knew the shock was giving way to the reality of two very serious breaks which were about to hurt like hell.

The ambulance arrived, along with a pick-up truck carrying nurses and a doctor. A Congolese nurse walked up to us, a picture of confidence and command. I gave as accurate a report as I could and handed over my patient to the professional. I returned to my truck and slowly drove around and back to camp. The scene disappeared in my rear-view mirror. The sun was still setting; the sky was completely lit up now with lavenders and deep violet, and far-off clouds the size of countries stacked high.

Today was a shift change. Bacon was being sent back to the UK with a couple of other staff. Our expat team shared a group of small houses, located in a camp around ten kilometres from the mine. The houses were small staccato buildings, nestled against green lawns and a tree line. After a shower and change of clothes, Bacon and I blew through the commissary for a quick bite and then headed on to the bar. It was a small, stand-alone white brick building with an open outside area with plastic tables and chairs. A few of the high-level chaps had beaten us there, and a small fire was already going. The bar inside was dark, so drinks in hand, we headed outside.

A couple of South Africans were complaining about

the troubles of getting materials into the country, while an animated and heavily bearded Spaniard from the Canary Islands kept piping up. He seemed to be agreeing but no one other than myself understood him. Bacon settled into his beer, and I was getting my Foreign Stout Guinness put away before starting on my G&T. I heard a familiar voice broadcasting from the far side of the field.

"Jez!"

James was striding across the lawn as only he could do. Coming in at something like six-foot-four, his t-shirt and shorts did nothing to hide the physique of the professional athlete. The son of Jamaican and English parents, raised in Spain, England, and France, he spoke a few languages and had spent his time in the DRC building on that. At home in the UK, he was a semi-pro rugby player, and soon after our stint in Africa, went on to be on the second string of the UK bobsled team. He was handsome in a way that caught the attention of women of any age, and his charm made it impossible to hate him for all the forementioned attributes. I still tried. He also insisted on calling me anything but my name.

We discussed his trip home, the ongoing site events and problems.

"Ah man," James reached over, hand on my arm. "Have you heard any news from outside camp? It's bad, Jez, we had serious security coming in." James was as happy a fellow as you could hope for, so concern from him needed to be taken seriously.

Reports of violence between local groups had surfaced, and the unrest was threatening to spill into our area. While we had security measures and evacuation plans in place, we were acutely aware of the tens of thousands of locals living nearby in villages and along the roadsides, and of how quickly the situation could descend into chaos and bloodshed.

In the West, we often view migration as a one-way flow. We see images of boats packed with refugees crossing the Mediterranean and hear the US news fixated on the southern border. In our narrow, self-centered perspective, migration is always a straight line that ends at our front doors. Human migration is, instead, a fluid and complex issue heavily shaped by climate trends that many have been late to acknowledge. Drought, floods, economic collapse—and the conflicts and wars that follow—have pushed millions from their homes, not only across borders, but deep into lands already lived in by others. These factors have a significant impact in Africa, and when combined with national, tribal, and linguistic tensions, the resulting pressures can be devastating.

Across southern Africa, countries have borne the burden of those fleeing conflict in northeastern Congo, but the greatest strain comes from the movement of people within the DRC itself. It was a migration decades in the making. Not far from our mine was a community of people from the Kivus. As it happened, locals became suspicious of the displaced newcomers and eventually saw them as a risk. Any local crimes were instantly blamed on them, and any effort to find work was seen as either stealing scarce jobs or driving down local wages.

Meanwhile, the refugees were trapped, living off meager aid provided by brave locals and aid groups. Tensions had been simmering for over a year, and what James witnessed was the spark that could set everything ablaze: large crowds of angry locals protesting and demanding the encampment's removal. All it would take was a single argument, a violent clash, or an act of desperation to ignite things. I thought of refugees I had seen elsewhere. There is no worse state for people than to be without hope. James was right to be worried.

We cheered Bacon farewell, and I reminded him to check in by phone when he had a signal so we could keep an eye on the security situation on the road to Lubumbashi.

Some weeks had passed since the initial worries of violence in the refugee community, and we had more direct matters at hand. A water pipeline, used to move surface water from the tailings facility back to the plant, skirted the property lines of the mine. Although it stayed within the lines, it impinged upon the edge of a farmer's field where maize was being grown. The farmer had planted more rows on the other side of the pipeline and now was complaining that it was within his field, and he had trouble getting around it to harvest. He was both right and wrong. The corn was impending onto the mine's land, as the farmer well knew, but the corn was no obstacle to our work. Our pipeline, however, was an obstacle to his crop. This was not the hot battle of wills it could have been. Both the farmer and project staff stayed cool-headed.

I had been dispatched by the mine with Xolani, a young South African engineer, to have a walk about and see the farmer's field. There were community communications people at the mine, but they were often overwhelmed, and our team would always try to help where we could. The red dirt road that demarcated the mine property was bordered on both sides by flowing green, grass, trees, and flowers: golden dewdrops and lantana popping up in colourful bunches. I heard some laughter behind us and turned to see a group of schoolchildren. They appeared to be between the ages of six and thirteen, a mix of girls and boys dressed in brightly coloured clothes. They carried their plastic sandals in their hands, their bare feet leaving tracks in the muddy patches along the wet road. As they drew nearer, their laughter grew louder; they whispered to one another before bursting into laughter, clearly at our expense. It did my heart good to see such carefree smiles—just a group of kids, like anywhere else, on their way to school without a trace of fear or worry.

The children walked past us, casting long glances back as they continued. They disappeared into the maize. Walking out of the tall stalks, like a character from a film, was the farmer. Joseph looked to be in his fifties, though he could have been younger. I recognized him from a group meeting held some weeks ago. Today was simply a walkthrough to explore potential solutions. He wore a light sweater and work pants, which in this heat seemed like madness, but for Joseph, this was a cool day in the mid-thirties. My polo was soaked through with sweat, the bright orange flagging vest made me feel even hotter. Joseph waved at us and walked along the road. My French was not up to the task, so Xolani did the translating.

Joseph gestured toward the small cornfield on the mine's side of the road. It was a triangular patch on an unused stretch of scrubland, but it encroached on an access road that also housed the large black pipeline—the true source of concern about the plot. Through Xolani, I explained that the corn could be allowed to grow and be harvested and that the mine would purchase not only the crop on its side but also three rows from Joseph's property, paying well above the local market rate. Food access for local communities was an issue, but they had wide access to vegetables. Meat was a rare commodity, fueling both the hunting and the trade of bushmeat in the region. This not only caused already endangered animals to be pushed further to the brink but also created pathways for disease. In a country already dealing with Ebola and possibly the Marburg virus, these were events that concerned international organizations as well as those in Congo.

Joseph seemed pleased, but none of this was news. This outcome had been discussed at previous meetings. We talked about how best to harvest and move out the corn while mine activity was occurring next to him, arranging days for farming and days for machines to work. I suggested adding his stalks and any other unused plant material to our live-soil stores. When we excavated an area, the mine would keep piles of very fertile, organic soil and plant material and save it for future use when we remediated an area the mine had been active in. It was becoming a common practice to "re-veg" an area that had been dug up so that the natural landscape could have a head start at regrowing and returning to its previous state. In a place as wildly fertile as the hills we were working in, this was less of a worry than in other places like the high Andes or a taiga forest. But being prepared and organized was always better than the alternative.

These ideas seemed to satisfy Joseph, and we agreed to keep talking. As Xolani and I walked away, I thought about farm, the half a dozen or so crops Joseph grew there, and how his proximity to the mine was both a blessing and a curse. The mine could provide jobs, a vibrant market for buying goods, but it also carried the weight of relentless operations, large machines churning up dust and noise, and the never-ending chess game of land rights and use. Mines also took up space, in all kinds of ways, from the actual mining activity and the structures required to the waste facilities, like the one we were building.

My mind then went back to the children walking to school, carrying their flip flops so they would not get dirty. What was their future? The balance between local communities and mines was fraught with challenges, often compounded by national governments' limited understanding of the needs, risks, or fears of either side. There was also the issue of history, both ancient and recent. Wars rarely bring stability or predictability, either for citizens or for companies. Conditions, expectations, and even laws can change fast. These were more snakes waiting in the grass.

Tunnel Between Worlds
Antamina, Peru, 2000

We were blasting a hole through the guts of a mountain.

High in the peaks of the Andes, under the cover of night, we were building what was known as a decant tunnel. Its purpose was to carry water from a network of surface canals through miles of solid rock and release it on the other side. This ensured that surface water—streams and rain runoff—was diverted safely away from both the tailings and any mining activity. The tunnel not only helped keep the water clean but also minimized the impact on nearby communities that depended on it. At Antamina, one such community was just a short walk from the edge of the mine's property; so close the tailings dam loomed large in their skyline.

This plan had another benefit: reducing the amount of water entering the tailings impoundment. While mines rely on water and recycle as much as possible, in the Andes, abundant rainfall—and occasional snow—means water scarcity isn't the issue. Instead, excess water can pose challenges, causing the impoundment, the reservoir behind the dam, to fill too quickly. Maintaining control of the water level is essential, ensuring there's always a safe buffer, or "freeboard" between the water's surface and the top of the dam. This freeboard is a critical factor in operational planning and monitoring, as it prevents water levels from rising dangerously high. A decent tunnel not only protects the surrounding environment but also helps ensure the dam's safety.

It felt strange to be working underground while perched at high altitude. As they say, the weather down there never changes. I was managing a team of contractors and engineers working on building the inside of the tunnel, and ensuring the work was done as designed and to a high level of quality. Our crew was a gritty mix of Peruvians, Chileans, and Colombians. They were led by Canadians, most of whom hailed from Sudbury deep in the Canadian Shield. They had cut their teeth as young men mining some of the planet's oldest rock and now traveled the globe. Their work always led them in the same direction: down.

The Canadians were easy to spot in their bright red overalls and hard hats. I'd head into the tunnel to watch them drill at the advancing face. Water seeped from cracks in freshly dug walls and pooled in puddles. The rock resembled a giant, three-dimensional jigsaw puzzle—grey-and-black angles intersecting with jagged joints. As the machines scraped, burrowed, and hauled, the pressure caused bits of ceiling to pop and ricochet off equipment like bullets. Running my hand along the wall as I walked, I could feel grit and dust of rock that had been buried for hundreds of millions of years, once part of an ancient seabed and pushed up over eons to form some of the world's highest peaks.

On the other side of the mountain, a second team was drilling from the opposite end. They used complex survey data and experience to ensure that when they intersected, the two tunnels would align. I would watch from the back as the two machines spun steel into the rockface, water spilling out and the noise screaming up the tunnel. The drill holes would be packed with explosives, and the crews would run blast lines back up the tunnel to a safe distance. After each blast, a thunderous crack shook the rock beneath our feet. If you looked up, you could see the shock wave shoot up the tunnel, bits of rock being knocked loose ahead of the pulse. After the blast, a crew would check to see that all the explosives had been set off and that there were no misfires or delayed charges. The crew would then go back in and, using jumbos—underground drilling machines—start excavating the rubble. Then, engineers and geologists would work the face, mapping and modelling the rock structures to gain a better idea of what kind of ground they were tunneling through, and what structures, if any, were required to make it stable for further drilling.

The process was repeated for months. Eventually, the crews met in the middle, with a group of guys waiting quietly while the opposing team chewed through the last short section, surfacing to cheers. Once the rock was stripped away, the matching tunnels were within five centimetres of each other's alignments. This outcome always astonishes me. Two teams trying to thread a needle under a mountain, in the dark.

Our oasis during these shifts was the sea can: a long, grey container fashioned into an office, stationed fifty metres from the tunnel portal. Inside were three small desks, a filing cabinet, a coffee machine, a bookshelf, and the smell of camp-cleaned clothes and fatigue. Near the door was the underground gear: head lamps and charging batteries, flashlights, and a stack of first aid kits. Muddy boot marks painted the floor.

Yuri sat quietly in his corner, reviewing the day shift's rock mapping after completing his first tour of the night. He was updating the geological classification model—a detailed map used to understand the types of rock the tunnel was cutting through. This model, built with complex software, provided critical insights for geological engineers and tunnel builders, identifying potential weak points and guiding the methods needed to safeguard both the tunnel and the surrounding rock from disaster. The data would then be sent to Canada for the tunnel designers and engineers to analyze. This cycle of review, update, and analysis would continue as we pushed deeper into the mountain.

Yuri was a Russian geologist who had been living in Canada for some time. As with many people from the rock team, he thrived on the technical riddles that surfaced almost daily. But his wide smile was absent tonight, so I gave him his space and spent the first part of the night grinding through my own problems. After a few hours, I looked over to see him sitting with his head in his hands.

"Headache, *brat*?" I asked him, using the Russian slang for brother.

I saw then that he was weeping. I poured us both another coffee and took a seat.

"What is going on man? You ok?" I asked.

For people who work in dangerous conditions, the emotional well-being of colleagues matters. A guy who just lost a parent or whose house just burned down may not make the best decisions in an emergency. Pulling them from the front line is often the best move. Yuri leaned back and wiped tears from his eyes. He gave that smile of those in the throes of loss.

"Sorry brother, my heart is too soft. I am ok, it is a tough day. They are going to leave them there. They are going to leave them at the bottom of the sea," he said in his deep accent.

I had been aware of the *Kursk* from the news the day before: a Russian sub that had been declared missing. It turned out that a torpedo mishap had escalated into an explosion which had blown through a main bulkhead and sunk the sub. The few survivors who had made it to a final, safe ninth chamber had succumbed to a terrible death by fire and oxygen deprivation. To make matters worse,

Russia's response had been a debacle, focusing more on managing the Kremlin's image than on saving the trapped sailors. By the time the British and Norwegian divers were permitted by Russia to attempt a rescue two days later, the sailors were all dead. Yuri and I spoke in the days between these stories.

"Putin vacations in Sochi while these boys sit and wait to die in the cold and the dark." Yuri was weeping again. I was struck by how connected he was to the event until he explained his brother was serving on a Russian sub. "Just boys," he kept repeating, his dust-covered hands shaking.

We spent the next three tours of the tunnel together. Over the months, I'd grown deeply fond of Yuri. His voice had a comforting warmth, like a heavy blanket, and his smile could wrap around you, making you feel that everything was fine. It pained me to see him struggling, and our quiet walks into the underground felt like the only way I could help. At the tunnel face—the wall marking the tunnel's current end—he stood, drills and equipment pulled back to give him space. He sketched the joints and structures in the rock, his lips moving faintly as he took both mental notes and meticulous entries in his book. Yuri was the rock whisperer. He read the earth like a tracker, piecing together the geology and geomechanics with a practiced eye.

Every time we exited the portal, I took off my hard hat and looked up to stars pouring across the narrow crescent between the two mountain peaks we stood between. This heavenly ceiling was countered by the shockingly bright lights of the work areas. Insects, even at these altitudes, swarmed around the lights like alien fighter jets in an interstellar dogfight.

Once the two ends met, an army of contractors took over to build out the tunnel's infrastructure. My partner, Zack, and I spent hours traversing the entire length of the tunnel. A Toronto native and a skilled engineer, Zack stood over six feet tall with blond hair and carried the innocent air of a farm boy. When he was in Peru, I was back in Canada, and vice versa. On our crossover days, we'd explore the underground together as he showed me the progress made during his shift.

As I pulled on my gear, the altitude was still hitting me hard. In mining, there's no easing into it, unlike climbing, where you adjust slowly. Here you're driven straight to 3,000, 4,000, even 5,000 metres, and expected to get to work. I developed a habit of chewing extra strength Tylenol like candy, trying to dull the sharp pain that gripped my head. We'd throw on our bright orange cruiser vests, stuffing notebooks and cameras into every compartment, and strap on the battery belt for our headlamps. Workers usually clipped their lamps onto the front of their hard hats, but engineers and managers had adopted a different habit. We'd sling our lights over our shoulders, holding them in one hand as we walked, using them to point things out. I say it gave us more freedom to highlight what we were examining. But if I'm honest, I also thought it just looked cooler.

Zack and I would enter from the upper portal, walking through the entryway and past materials stacked near the entrance. We would make our way past the deep shaft on our right where the water was going to drop down from the canals along the mountain ridge above us and then, when we were finished, through the tunnel and out the other side of the mountain. This would bypass the natural flow from several streams and rainfall around the mine and tailings facilities, and out and back down the valley we were interrupting with our structures.

We walked along the tunnel, a hundred metres of concrete-covered walls and ceilings. On the surface was information to decode: cracks and blemishes, perhaps some small gaps along joints. Zack and I would look them over, take our notes, and look for signs of trouble or distress. Materials talk to you; they tell you what is going on. I would run my hands along the concrete, cool and wet to the touch, pulling a bit at the edges of a small crack: Was it deep or only superficial? What shape did it take? Was it just a few centimetres or did it go far?

We ventured further. The tunnel, lit up with working lights, was eight metres wide and roughly twelve metres high at the top of the arch. We moved past people and machines and stepped up onto the platforms, climbing higher and higher up into the maze of steel and wood, wiring and hoses. Waterfalls of sparks dropped down on us from above as we continued, men cutting steel along the ceiling, installing braces against the rough rock.

Along the top of the tunnel, mounted on the ceiling, was what looked like a large, brown vacuum hose. This was part of the ventilation system that kept air moving underground. Fresh air was force blown in and the action helped in pushing the build-up of stale air and gases out of the tunnel. The air quality was monitored several times a day, not only for gases but also for particulate that came from blasting and general works.

Zack and I climbed our way through the scaffolding, around pumps and small groups of men drilling anchors into the rock walls. You could smell ozone and fire, diesel exhaust and dust burning on hot lights. Every so often, we would stop again to look more closely at something, some rock joints where water was coming in, a pile of tarps that was stacked too close to an ignition source. We would make our notes, talk to the supervisors and workers; we would take photos and ask questions of each other. And then, continue walking.

Eventually, we made it all the way through and out to the lower portal, on the other side of the mountain. An entire camp of people and equipment was working at this end. The night air had turned sharply cooler. My breath drifted upward, caught in the glare of a dozen floodlights trained on the worksite. Zack and I turned around. The journey in this direction had been entirely downhill, but now we were to battle that hill the whole way back. The tunnel was roughly 1.3 kilometres in length, and it was normal to need to walk it two or three times a night. On a few nights, I counted five trips in one shift. During one thirty-day rotation, I lost a total of twenty-seven pounds.

As we walked back, it occurred to me that this was as far away from civilization as I would likely be in my life. We were underground, at high altitude, on night shift. As we struggled up the tunnel, my breathing becomes heavier as the grade and lack of oxygen began to tear at my lungs: my boots were as heavy as iron, my legs wobbly.

I looked up ahead, seeing the lights shine against the ongoing works, seeing the gaps between forms and the rock walls. We stood in absolute shadow, out of reach, our headlamps off. I felt completely removed from everything. I thought of the Greek underworld of Tartarus, a place so shrouded that, in Hesiod's words, "night is poured around it in three rows like a collar around the neck." I watched night pour around us, section by section. Somewhere down here, Cronus wandered, king of the darkness.

How to Kill a River
Lima, Peru, 2024

THE CITY OF LIMA RESTS at the end of an alluvial fan over 400 kilometres long that drops down from the town of Ate at the base of the Huaycoloro Basin, all the way to the beaches of Mira Flores. It is a long downhill ride that carries you through the lower foothills, with multiple peaks of rock sticking up through the sands and gravels of the fan, and then the lower coastal hills which seem to be no more than giant piles of sand. Upon these hills sit communities of thousands, small hand-built dwellings stitched into the hillsides like a multi-coloured quilt of plastic and wood.

It is along this fan that the river flows, from its headwaters, cutting a valley through this ancient structure of sand and gravel and crumbling mountain slopes. But the river does not flow as it used to, and due to droughts, water use, and diversion along hundreds of communities, it is running at an all-time low.

In Peru, as in many places, there is a narrative regarding the abuse of water sources by large mines and mining companies. A large mine in operation can use immense amounts of water; one of the major factors in determining whether a mining project is even feasible is the location's access to water. The people of Peru are no strangers to extraction industry. Much of their history is intertwined with mining: from the ancient civilizations of the Inca to the Spanish. Indeed, it was the gold that the Spanish encountered that drove their thirst for land and conquest.

But is this abuse of water sources true? Make no mistake: mining's legacy is one of deep and enduring environmental harm. But, as with many things, the modern narrative is shaped as much by hyperbole as by facts and conditions. These narratives can be effective in building consensus for movements and campaigns but can fail to consider the actual situation on the ground. Indeed, the difference between what really happens on a mining site and what people believe is happening can be immense—and this goes beyond the true environmental cost of any single operation. After speaking to hundreds of students, community leaders, and concerned citizens in various countries, I've come to realize most people have only a vague sense of where the materials that sustain their lives actually come from, or what it takes to extract them.

Let's go back to water. Ask the mayor of a city or town about their community's daily usage, you'd likely get a puzzled look. They might mention an average per household or toss out a rainfall statistic, hinting at the shifts brought about by climate change. Ask a mine manager, and you will get a very different answer. They will provide a precise figure, likely measured in cubic metres. From there, they would explain the details of reuse—how much water is recirculated across different systems. They would outline usage in the camp where workers live and eat, in the plant where metals or minerals are processed, and in the tailings transport system, where water pumps waste materials through pipes to the storage facility. Consumption data would be clear, accurate, up-to-the-minute.

It has often struck me that cities could benefit from adopting a similar system. Communities often seem clueless about exactly where their water comes from, where it goes, where it's wasted, and how it can be used more efficiently. Large cities and towns have extensive civic departments managing these tasks, but there's often a significant disconnect between local leadership and residents. Of course, practice follows need, and some communities fare better than others. My friends in Southern California can tell me precisely what their allowable water usage is. Meanwhile, when I ask my sister in Seattle, she looks at me like I'm speaking a foreign language.

There's no question that Peruvian towns have suffered the brunt of mining projects, an example being the Doe Run project in Huancavelica. The project was made famous by the successive mines and owners creating a perfect storm of contamination, which led to the town of La Roya being declared the most polluted place on earth.

But when you take this history and these examples and

hold them up to modern-day mining practices, the situation becomes far more complex. The mines present large, visible impacts on the land, but many of those impacts—tailings dams, surface water canals, pumping stations, fences, and roads—are the very structures and systems reducing and mitigating the dangers of years past. At the same time, on every project I have worked on, great pain is taken to reroute waterways such as rivers and streams around the project and back into the environment. Projects constantly monitor the quality of the water around them, both on the surface and in the ground. This is where the narrative of what is happening takes on a third dimension, one that conflicts with prevailing public outrage.

Across the hills and valleys of these regions, artisanal miners are a common sight. These small crews work on a modest scale, extracting minute amounts of gold, silver, copper, and other minerals from the rock. The process is fraught with physical danger and relies on highly toxic methods. Without the oversight or resources of a company—let alone proper health and safety protocols—miners use bare hands and unprotected lungs to handle hazardous substances like mercury, creating a chemical amalgam to separate valuable materials from the waste. And the waste? Barrels and buckets of toxic compounds and debris are often dumped straight into the rivers. And yet, I have watched these same groups of men and women attend public meetings or join protests to contest what they see as the mining company's destruction of the land they live on.

This isn't about assigning blame but about recognizing that solutions require a deeper understanding of what's happening and how we can mitigate, reduce, and eventually eliminate the impacts—while also building systems that protect both the land and its people. Many educational programs have been launched to highlight the dangers of these activities, along with job creation initiatives aimed at providing safer alternatives. It's clear that no one willingly chooses the dangerous and backbreaking work of small-scale artisanal mining. This is a last resort for those abandoned by the failures of the prevailing economy.

In some regions, such as Colombia, Mexico, and parts of Africa like the DRC, organized crime forces coerce people into these roles. In South Africa, small-scale wars erupt between rival gangs fighting for control over illegal mining operations and sometimes even on active mine sites. In the DRC, child slaves, often abducted after their villages are raided, make up the primary labour force in many artisanal mines. These grim realities demand our attention and a collective effort to end the suffering inflicted on both the people and the environment

There is more to this issue of water, and in Peru, it is a question that follows the Rímac River down from the high glaciers and slopes of the Cordillera Negra down toward the sea. When I make these journeys, the highway snakes across steep cliffs before plunging from the high plateau into deep cuts carved into the earth. As I descend from the highlands to the shaded valleys below, I think about this open wound—this gash in the landscape, shaped by millions of years of water, wind, and the restless movement of the planet itself. And I can't help but see the parallel: these natural scars are not so different from the ones we create with our explosives and machines, tearing into the land to take its minerals and metals.

The road begins to thread through an increasingly dense population. Small villages give way to crowded towns, where traffic chokes to a crawl as buses idle mid-lane to load passengers and trucks spill their cargo onto the roadside. The rhythm becomes stop-and-go, then mostly stop, especially as Lima draws near. The countryside falls away, replaced by the sprawl of low-slung buildings and corrugated tin.

As you drive in, the river comes in and out of view. Bright, striking graffiti appears on high walls and buildings. Bridges cross overhead, smaller tributaries feed in, and man-made channels direct runoff toward the water. Along the way, one measure of the population stands out unmistakably: the plastic waste accumulating along the banks and floating in the river. Several bridges mark elevation drops where the Rímac tumbles over concrete steps or natural falls into lower pools before continuing. At these junctions, plastic and bottles pile up, forming debris dams; towering heaps that, every so often, break loose. When they do, an observer can watch the slow-motion collapse, an avalanche that carries with it everything imaginable: tires, diapers, oil drums, toys, clothing—anything you'd find in a municipal landfill flows along with the Rímac.

And these are the things that can be seen with the eye, but the real danger lurks out of sight, heavy metals and bacteria, viruses and waterborne disease. While you drive, you can see flashes, oases where the river almost looks restored. Bands and pools that have perhaps been cleaned out, by man, rains or a sudden flood. You can look down and see people washing clothing by the riverbanks, children playing in the water. It would be a serene and peaceful scene if not for what you passed earlier. The knowledge of it weighing on you makes you want to cry out to the children to get out of the water, like a scene from *Jaws*.

I have stood overlooking the banks of the Rímac on an early morning in the Santa Maria de Huachipa municipality just outside Lima. It was early morning as I watched trucks back down a slope and dump refuse straight into the river; the very river city water is sourced from. Just a few hundred metres downstream, I have watched people picking through the waste that line the banks. It was no different from the garbage pickers I had seen in Congo, Bolivia, or Nigeria: young women carrying babies on their backs, next to huge plastic weave bags into which they throw what items of value they can find amongst the mountains of detritus.

A question I ask is: Why? Why these piles? Why the dumping? And the answers are, again, both complex and simple. The city of Lima, especially the outer municipalities within it, has either no or very poorly organized garbage management. What I witnessed was the combination of failures on every level: education, governmental, law enforcement, community pride, and engagement. People who feel connected to their home do not dump garbage in it.

The surge in plastic use, however, also mirrored an explosion in population, driven in part by mass migration from the countryside to the city. Lima has long been Peru's economic and cultural heart, dating back to the Spanish colonial era. But during the turmoil of the 1980s and nineties, when terrorism and near civil war gripped the country, the capital became a refuge for those fleeing violence from both insurgent groups and government forces. During this period, millions of rural Peruvians poured into Lima, swelling its neighborhoods. Even before this, Peru's wealth of resources had made it a destination for immigrants from China, Japan, and other Latin American countries. The result was not just overcrowding in existing districts but a rapid expansion of urban sprawl stretching into valleys toward the Andes and spreading up and down the coast. That expansion continues today, visible in the low-cost developments I mentioned earlier—grids of streets mapped onto the sand, waiting to be filled.

These migrations coincided with a surge in plastic use for food packaging and everyday materials. In the 1990s, there was even a campaign promoting plastic as a replacement for paper products. It was a strategy meant to curb deforestation and protect the Amazon, among other vulnerable forests. I remember those campaigns. I remember the songs about saving the rainforest. These were well-intentioned, necessary movements.

But now, as I watch plastic pile up along the banks

of the Amazon, the Niger, the Fraser, and the Rímac, the old promises of recycling and sustainability echo back like faded slogans. And as I see trucks dumping municipal waste straight into the Rímac's banks, I can't help but wonder: What's the fix now?

Make no mistake, mining impacts the earth and the environment around it. My goal here is to talk about what I have seen, and to ask questions—questions for which I do not have answers.

I feel a disconnect between the people of Lima, their justified anger about the abuse of their land and its resources, and the actual sources of that abuse. I remember one evening in a taxi, passing a pile of garbage on a street corner, when the driver shook his head and muttered, "Ah how ugly. Lima could be such a beautiful city, but people act like it is a toilet." I understood his frustration completely. And yet, two blocks later, he casually tossed an empty glass Coke bottle out the window. I heard it shatter on the pavement, but he showed no reaction. It wasn't satire or a statement. He simply didn't connect his action to the problem he had just condemned.

How do we build that connection? And not just in Peru. Rivers around the world are under strain. In the southwestern United States and across Europe, water levels have been steadily falling. The great rivers of India are becoming highways for disease. And in Bangladesh, the ancient deltas turn against the people every monsoon season, pushing millions from relative poverty into displacement. The 1970s saw a dramatic reversal in river pollution in North America, thanks to the creation of the EPA and other regulatory agencies. But in much of the developing world, the arc bent in the other direction: the same companies that cleaned up at home moved their factories abroad, exporting pollution to countries with weaker laws and fewer choices.

Extend these conditions outward—a warming climate, a growing population, accelerating migration—and the cracks in the world's foundational systems begin to show. Layer on the failures we've already seen, such the power grid collapse in Texas, or gas shortages rippling across Western Europe, and a pattern emerges. It's tempting, from a place like Canada, to view the hardship in Lima with distance or pity. But the truth is harder, and closer: what we're seeing there isn't an outlier. It's the edge of a future already arriving. Not a cautionary tale. A preview.

Checkpoint Lives
Antioquia, Colombia, 2019

The gold mine sat in a valley so deep it erased the sky. Halfway up, two plateaus emerged. The top one held some housing and the commissary; the second, further down, was lined with micro-bungalows made from stand-alone containers. Some staff preferred the housing above, older but with larger footprint and a short walk to chow. I liked the smaller, newer spaces, and the privacy they provided. From the lower ledge, you could watch clouds cruise past the tree-covered slopes or see thunderheads drop their downpour like fire bombers.

The mine never slept. As tailings area manager, I was up before dawn, often before three, driving the steep, twisting descent to the bottom. Rain lashed the windshield and lightning burst like blue tubes of light just beyond the truck's edge. From the higher roads, I could see the lights of thevillage, pressed up against the mine. I drove through the entire property, past foundations and buildings beginning to take form, past the office complex cobbled from stacked shipping containers—the makeshift nerve center where meetings happened, and reports churned endlessly. I descended past contractor buildings and workshops along the right abutment of the valley, stockpiles of material and the excavation for the tailings stack on the left. Parked dump trucks lined the route, some idle, others roaring awake with black smoke billowing from their tailpipes, like early morning dragon breath.

Along this stretch, I passed the tailings impoundment where the ground was being prepped for black plastic liner. Workers rolled it out, cut it to fit, and welded it together in thousands of seams and patches. By midday, the heat turned the surface into an oven, the plastic covered in boot prints, scattered strips of cut liner, and a dense network of markings: numbers, arrows, test codes, and the initials of welders. These were the runes of installation, a cryptic shorthand that mapped the story of the liner's construction. This was where the tailings—a dry expanse of grey gravel—would be piled, compacted, and stacked ever higher, filling the valley; a moonscape emerging from the jungle floor.

At the bottom of the site, past the security gate, the valley tapered into a dirt path. I'd walk the perimeter in the dark, checking for damage—rain undermining the work, vandals testing the boundaries. Then back to the container offices, an hour for reports and messages before the morning manager's meeting began. The machine kept going. The valley kept filling. The work pressed on.

On a cool Tuesday morning, I headed back up the hill with Manny, part of the group overseeing all the work contracts and payments for the work being done. We passed a squad of Colombian Army cadets on a march. They trained by trudging up and down the steep mountains, full packs strapped to their backs. These were boy soldiers; fresh-faced, still being broken in. They were marching through a valley that, not long ago, had been a hub for the Medellín cartel. They moved uphill in small, deliberate steps. Their rifles, factory-new, stood out—some still bore stickers from Colombian manufacturing plants. As we passed them, the soldiers grinned through the pain, sweat darkening their green hats and shirts.

We climbed higher, passing buildings and laydown areas along the steep ascent until we reached the checkpoint at the mine's entrance. There, security guards outfitted in knee pads and Navy SEAL-style helmets waved us through. They took their job seriously—maybe too seriously—and were something of a running joke among the miners, meticulously checking behind seats for stolen contraband or anyone trying to sneak alcohol or drugs into camp.

Once through, we were on paved roads, a winding one-and-a-half lane that ran past homes built on the very cusp of the cliffs, brick structures that lined the road with women and children sitting in front doorways, frighteningly close to traffic and equipment. Toddlers would stare as we drove past, sitting still in what was sometimes crushing heat. We wound our way up toward the main highway, Route 62, and here came to the real checkpoint.

Gone were the fresh-faced recruits on their first deployment. These were men in their late twenties and thirties, hardened, disciplined, and unmistakably professional. After years of what was essentially a civil war waged by overlapping factions across shifting territories, Colombia developed security forces that operate in ways different to other nations. Here special police groups handle high-stakes counterinsurgency, carrying out special operations against narco and paramilitary groups. They carry well-worn AR-15s, and light machine guns with faded metal, always in hand, never slung over a shoulder. Their emergency medical kits were strapped front and center, their boots had seen miles, and their eyes didn't scan truck beds or back seats. They met yours directly, looking through you, not past you.

I've seen professional soldiers across the world, from Northern Ireland to Angola, and these men had the same air about them. Police or not, they were the ones on the front line. Our stops at their checkpoint were never more than a few seconds—a quick question, a wave through. They weren't searching for stolen tools or contraband beer; they were looking for those who planted bombs and set ambushes. And I was glad they were there.

A few months earlier, these men had flagged down a box truck on the highway, a routine stop that quickly turned into something else. The driver and passenger had a look about them; something was off, something that made the guards' instincts flare. The moment a hand went up to halt them, both men bolted, abandoning the truck where it stood. When the squad pried open the cargo hold, they found it crammed with weapons, ammunition, and heavy ordnance, including grenade launchers. Intelligence later revealed that a paramilitary group had been preparing to cut off the region, erecting armed roadblocks and taking hostages. That single stop didn't just foil their plan. It set off a wave of arrests across Antioquia. And all of it had happened right where we were stopped.

We passed through the check point and, about a kilometre up, pulled into a truck wash station run by a local family. It was the only place we could keep our trucks remotely clean. We set up at an outside table while Jose, the station owner, began blasting off layers of mud. As we sipped Fantas, we talked business—mine traffic, security. Jose told us about a couple of kidnappings further north involving a road crew. He sang to himself as he worked, singing against the flow of the music that we could hear coming from a speaker somewhere out of sight. Jose suffered from some physical malady, which I could not name. From his slightly disfigured face and hands, I assumed it could be leprosy. He moved quickly, his bare feet splashing in the puddles; his feet seeming to have avoided the ravages of the disease.

After a short while, three women emerged from the building on the right and began hanging laundry on lines strung from the wall to a pole near where we stood. A few minutes later, one of them disappeared inside and returned carrying an enormous chicken, resting calmly in her arms. The bird was strikingly coloured, a length of yarn tied around one ankle. I watched as she reached for a long-handled broom leaning against the wall. At that moment, the chicken, which had been perfectly content, grew agitated. Sensing the shift, she adjusted her grip and with a practiced motion, flipped the bird upside down, grasping it by its legs. The chicken squawked and flailed as the woman laid the broom on the ground. Her face never changed; to her, the flailing bird might as well have been a sack of grain.

She propped up one end of the handle with her foot. Then, in a swift, deliberate movement, slid the bird's head and neck underneath. The fight stopped instantly. The chicken froze, hypnotized. The woman placed both feet firmly on either end of the broomstick. Then, with a sharp motion, she twisted the bird's body by its feet—once, twice. On the third turn, the head detached cleanly. She remained still, holding the bird as its wings flapped wildly, blood spilling freely from the open neck. The severed head lay somewhere on the ground, unseen but unmistakably there. A few loose feathers floated into the air, and then, as if a switch had been flipped, the body went limp. Without hesitation, she stepped to the wall and drained the rest of the blood over the hillside, as she had probably done hundreds of times before.

The sheer efficiency of what I had just witnessed left me mute. The bird's dispatching was no different than turning off a tap or taking off a shoe. Manny managed a quiet "Jesus Christ." I had to remember that he was a city boy from

Bogota and was possibly as far removed from these rough and tumble lives as I was. The woman gave the chicken a final swing, flicking out the last of the blood, and walked back into her kitchen. Jose whistled to us: our truck was ready.

When I arrived in Colombia, I immediately sensed tension within the team. The pressure to deliver was intense. Every deadline loomed large, and both schedule and budget were under constant scrutiny. After the first few weeks, however, I noticed something I had not seen on other projects: a terrifying rate of attrition. In the two weeks after I started, two senior managers were fired and replaced. After the first three months, nearly the entire electrical engineering team had been cut, with no replacements brought in. I had worked on lots of fast-moving projects, but the speed at which people were axed, and the circumstances surrounding it, seemed to border on homicidal. Even Larry, the American who had hired me on and was the project manager, was dismissed one afternoon over the phone from his home in the US where he was on days off.

I realized we were all just one serious problem away from being canned. Once that clicked, my anxiety about getting fired faded. Sooner or later, I'd get the call—asked to stay behind after the morning meeting, told the truck outside was waiting for me, thanked for my efforts, and given a day to pack up before flying out of Medellín. It was inevitable. And knowing that freed me. I stopped worrying and just got on with the job.

It was a complicated crew. Highly skilled, with centuries of combined experience under one roof. But beyond competence, there was a shared thread of identity that connected most of the team. This was 2019 and Trump was a very much celebrated leader here. With that, came a marvellous disconnection from reality. One afternoon, Brent, the crane lead from eastern Washington state, was having a meltdown in the office. He was raging against something he had read in the paper about new research findings, loudly declaring that this "liberal science bullshit" had hit its peak. He had no time for science or any of the lies that came with it. Americans, he proclaimed to the room, should just get a job and forget all this "fancy research crap." Not long after, I watched him settle down, open a textbook on dynamic loads, do a bit of analysis on a microchip he needed to replace in one of the newly arrived cranes, and seamlessly switch between his cell phone and laptop to get the job done. All the while, he was sweating through the bandana permanently tied across his head, more from rage than heat. I headed out to do another site tour. I suspect that at no point since that day has Brent ever connected his argument with the work he was doing.

We were working at an operational momentum that was difficult to maintain, and the strain was showing. Guys were starting at five or six in the morning, forever trying to catch up with impossible workloads and timelines and staying late into the night. Adding to the stress, every few days I'd get a midday summons to the office with little notice. Someone was arriving in an hour—maybe a prospective buyer, an investor, or a Colombian politician. There was always the off chance the country's president would show up. We never knew who it would be until we saw the helicopter banking sharply across the valley, followed by a convoy of tinted SUVs snaking down the road, perfectly timed with the landing. Each time, we'd gather at the lookout platform, an open area designed for these kinds of visits.

The problem was that while the lookout offered a perfect vantage point over the site, it was also a sniper's dream—should the visitor be worthy enough to clip. The moment a VIP was inbound, the area swarmed with security contractors and soldiers, radios crackling, rifles up. Once the convoy arrived, we'd go through the usual routine. Each of us, as area managers, would give technical briefings. A few questions, some photos, plenty of handshakes. Sometimes it meant a site walk, a tour of this or that. But more often, it was just a quick photo op before they were whisked back to their helicopter, lifting off in a swirl of dust and noise. Afterward, the security detail would grab a quick meal in the mess hall, then head out for the long drive back to Medellín. This happened about twice a week, piling yet another layer of strain onto a team already running on little sleep and frayed nerves.

Some people broke under the stress. I remember a very bright guy from Texas who had been doing an excellent job on the mechanical side of the operations.

He had been diligent and very easy to be around; funny, quick with a burn or a compliment. I'll call him Mike. About four months into my project, he was called out to a port city to carry out a hard inspection of some large and expensive equipment that had just arrived from Europe. I saw him off the morning he left and could sense something different, a sort of quiet dread. I asked him if he was good, and Mike just smiled and patted me on the shoulder.

"Yeah man, I will see you around, ok?" he said. It seemed odd. Final.

The next day, I realized Mike hadn't emailed or checked in as I'd asked. Our travel security company hadn't heard from him either—a red flag in a country where kidnapping, whether for political leverage or ransom, was still a real threat. Mitch, our offshore and regional security guy, expressed concern when I told him Mike had not buzzed me. "I will chase him down, see what's up," Mitch said.

Mitch called the hotel, and, to our relief, we were told he was in his room. He had dinner sent up and all seemed well. It was still strange to me that he had failed to contact us, but sometimes guys get busy.

Two days later, Mike missed his meeting at the port to inspect the materials. This was a meeting that had been difficult to arrange; a series of people around the world were waiting for the results of his audit. He had not checked in and had not emailed. The hotel once again assured us that he hadn't left the premises—that the cleaning staff had seen him that morning and delivered fresh towels to his room. Mitch told me he sensed something amiss with the hotel manager, something in his voice seemed wary. That afternoon, after a second no-show at the port, where now several people from the company headquarters in Medellín were in attendance, the situation needed action. Mitch called the hotel once more and demanded to speak to Mike. I watched his face; I could hear the Spanish of the hotel manager on the line. Mitch's face was locked in a frown. I wondered if that was the same expression he wore while peering through a rifle scope as a Scout Sniper in Iraq. Mitch hung up.

"Ok, he's alive, but we have a problem. It isn't good." The next day, Mitch called me from the hotel. Mike had arrived on schedule, walked in with his bags and a few from a local store, shut the door to his room—and then spent the next two days drinking enough hard liquor to kill an elephant. The news was devastating. Mitch spent two days in the hotel drying Mike out. He had a doctor come to check that Mike was not poisoned or on the brink of a medical collapse. It sounded like Mike was on the precipice, and we were lucky to have Mitch there.

We got word that Mike was no longer attached to the project. The standard offers of help and treatment were offered, but I think, having operated at such a high level, the embarrassment was too much. I was sad to know he would not be back. I sent an email that was never returned. Mitch took Mike to the airport, reminding him that he was the company's responsibility until he walked through his front door, and that he needed to stay sober until he got there.

As Mike boarded, Mitch told me, he glanced back, yellow-faced and shattered. In my mind, I pictured him shuffling toward the ramp, his olive-green shoulder bag dragging behind him, head down, worn to the bone.

Jenga
Puerto Balboa, Panama, 2002

Morning traffic in downtown Panama City churned with buses, taxis, and horse-drawn carts. Vendors wandered between cars, selling fruit and candy, housewares and straw hats. They liked to stand and chat through your open window while the flow was at a standstill. I often kept a pack of cigarettes or a bottle of water for them, and candy for any kids they had in tow.

Eventually, I broke free and headed south along the ocean. The road narrowed, and traffic thinned as I turned off toward Porto Balboa. Soon, I was among container trucks going in both directions: pick-ups with equipment and men sitting high in the back, workers sipping from small drink boxes or coffee from tin cups. Like any large project, the Panama Canal had become a gravitational force, generating its own orbit of commerce: jobs circling jobs circling jobs. Gas stations and truck stops appeared between the open green of the roadside. Small box trucks and pick-ups loaded with fruit and supplies. Papaya and bottles of water hung from strings and shelves of small wooden stands. Torregitas and hojaldras, pastries and friend corn flatbreads, were the morning staple along with bitter roadside coffee piled with sugar. You could smell the open fires. Music blared from small radios held aloft by passengers of motorcycles darting like sparrows between trucks and cars. When the arches of the Bridge of the Americas came into view, I knew the port was close.

I crossed the security gates. The port was slowly waking up. Workers milled around, short-sleeved men and women checking shipping containers. I drove to the far end, where the wide concrete expanse met the sea. I parked in front of a large warehouse with steel walls and a sloped roof. The building was nearly empty except for a space in one far corner: tables, shelves, and rows of low metal racks to stack samples on. My workstation.

I was in Panama to run a soils laboratory for a drilling program. The port owners, a Danish shipping conglomerate, wanted to build an extension on their existing operations at the mouth of the canal, and we had been hired to conduct drilling and geophysical surveys to establish what kind of soil conditions lay below the water of the bay. Our data would guide engineers in designing the foundation for the port expansion to come. Towering cranes would rise from reinforced concrete and steel platforms, anchored by piles driven below the seabed. This groundwork needed to withstand the test of time, enabling the global flow of materials through the canal and beyond for decades to come.

The drillers arrived in their small pickup truck, a near-derelict from our equipment supplier who was convinced he could save us money on a rental. Instead, he produced a rusty, dented Nissan which looked as if it should be pulled by a couple of horses. The older driller, who I'll call Leo, was a pot-bellied old hand from Ontario. His younger helper was a twenty-someone I will call Ryan. Leo opened the conversation, lamenting the lack of Timmie's coffee in Panama. After a brief planning meeting, the boys boarded their small boat out to the drill rig floating on the bay. The port was vast enough for our crew to operate without crossing paths with the endless stream of container ships and other vessels slipping through the canal. But as a young lab tech, I rarely caught sight of that waterborne world. I was buried in dirt.

This morning, I was extruding Shelby tubes—using a hydraulic cylinder to gently press soil from steel casings. These samples came from deep beneath the bay floor and would undergo a series of tests and assessments. Once extruded, each sample was labeled and photographed. Those requiring more advanced analysis were carefully wrapped and packed for their long journey to a specialized laboratory in Canada.

Puttering away in the shaded but oven-hot warehouse, I thought about the adjacent body of water and how it fit into the economy of the world. The Panama Canal was completed in 1914, the result of a century of effort and the deaths of hundreds of workers—lost to disease, heat, exhaustion, and construction accidents. Its opening, some declared, marked the birth of the modern world. What struck me most about this marvel wasn't just its scale or ambition, but the ongoing, almost invisible labour required to keep it alive. The canal demands constant attention: dredging the natural lakes that connect the engineered sections, using enormous machinery to scoop and vacuum sediment from the shipping lanes. That material is then hauled by barge, either dumped or reused, depending on its quality.

I know the people who do this kind of work—who build and operate the systems that keep the world moving. I've been one of them, in one form or another, for my entire adult life. Sometimes I found myself and my team at the centre of things, making decisions and steering the project. Other times, we were just a cog in the great machine—doing our part, then moving on. Here in Panama, it was clear: we were only one small part. It would be years before plans were finalized, contracts awarded, and the first machines arrived to break ground on the expansion.

Coffee in hand, I stepped onto the concrete plinth that surrounded the warehouse, the port and bay open wide before me. To my left, shipping containers were stacked high along the dock, behind blue cranes that pierced low-hanging morning clouds. A blue ship entered the bay, swinging in a slow, wide circle as it turned its port side to the port, where the cranes awaited its cargo. Tugboats were dwarfed by the steel monolith they pushed toward the dock. Someone else might have seen only a ship on the water. I saw more. I saw the threads tying this moment to thousands of other places—and, perhaps, millions of other people.

A Panamax ship can carry about 52,500 deadweight tons—the maximum cargo weight it can hold without sinking too low to clear the locks. (Without the canal's restrictions, the same ship can handle 60,000 to 80,000 tons.) The ship itself is a mountain built from materials: steel, copper wiring, pipes, miles of rubber for hydraulic lines and millions of parts made of plastic, glass, and wood. Now add fuel, various oils and coolants, foods and materials for the crew—and paint, tons of it. Not just on the hull, but over every inch of metal: seals, gaskets, connectors, even deep inside the engine room. It isn't just for show; it's armour against corrosion.

Now, reverse engineer each of those materials, one at a time. Copper from Chile and Peru, iron ore for steel fabrication from Minas Gerais, Brazil. Molybdenum, a material for hardening iron ore into steel, maybe from Kalkaroo in South Australia. The paint and fuel trace their origins to petroleum extraction, but also to the immense infrastructure required to refine and move that oil, whether from the deserts of Saudi Arabia or the waters off Venezuela. Transport itself is no minor player in this chain; the fuel powering the ship I'm watching almost certainly arrived aboard another vessel, or via pipeline or long-haul truck from places like Alberta. Each step is powered by resources that are extracted in turn.

It's almost too much to hold in your head at once: the endless entanglement of things we use and the systems that sustain them. Is this towering scaffold of interdependence just a high-stakes game of Jenga? Remove one piece—copper, steel, oil—and the entire structure might wobble and fall. We walk into grocery stores, order shoes online, boil water for tea, unaware that every gesture leans on a buried chain of labour and logistics.

I watch the ship dock. In moments, the cranes will pluck containers from the deck and set them down onto flatbeds. Pelicans skim the water, in search of fish. I think about the coming changes—the docks, concrete and steel—and how the contours of the bay will change forever.

Back at work, I carefully trim a small puck of soil taken from the seabed, fitting it gently into a steel ring so that I can apply weight with a strange, medieval-looking apparatus over a scale to conduct what is called a consolidation test. Pressure is applied until the puck relents. This test tells future designers what the resistance of this soil is. A graph I create here, and email to the other side of the world, adds a tiny clue to the greater picture of how this project will be built. The micro and the macro: me bent over a small dial, writing down numbers as soil gets squished into a frame, while nearby hundreds of containers are being offloaded from a ship on the edge of the canal.

As twilight deepens over Balboa and the sky turns a dusky violet, I finish packing my samples and equipment. The warehouse is quiet now, long emptied; Leo had left hours earlier after dropping off tubes and soil bags. I lock the door and head toward my truck. Out on the bay, another ship glides past the dock lights, bound not for shore but for the open locks. In eight hours, it will have traversed the length of the canal. By first light, it will be somewhere beyond the isthmus, sailing into the warm swell of the Caribbean. And from there, who knows? Maybe New Orleans. Maybe Rotterdam. Maybe Lagos.

Mitigation
Minas Gerais, Brazil, 2025

The forest floor crackles beneath my boots as I walk between eucalyptus trees and termite mounds. I've left my truck off the mine road and am picking my way uphill, chasing a vantage point over the valley below. The trees are a rainbow of inky blues and weird greens, the skin peeling in waves.

I'm in Minas Gerais, a state in southeastern Brazil. Its name—meaning "General Mines"—isn't just a label; it's a story. In the seventeenth and eighteenth centuries, this region was the epicenter of the country's gold rush. Today, the gold has dwindled, but Minas Gerais remains a mining powerhouse, producing, among other things, iron ore. But the people and the land have paid dearly for it. Less than a kilometre away, in the next valley over, lies the site of the 2015 Samarco Dam collapse: a disaster that unfolded in the dead of night. More than 40 million cubic metres of mine tailings and dam debris heaved through the towns of Bento Rodrigues and Paracatu do Baixo, leaving nineteen dead and contaminating 670 kilometres of river basin. Eight years later, the full extent of the damage remains unknown.

Then came an even greater tragedy. In 2019, the Brumadinho dam failed catastrophically, killing 270 people. It engulfed the mine's administrative area, including a cafeteria during lunchtime, as well as nearby homes. This is what finally prompted my team's arrival. We are among several international groups tasked with inspecting the dams and tailings facilities run by Vale—the world's largest iron ore producer, and the majority or sole owner of both the Samarco and Brumadinho sites. Its entire project list in Brazil was now under a microscope, and our role was to be the lens of that scrutiny: monitoring and assessing the work of decommissioning, or recharacterizing, other dams in Vale's catalog deemed "at risk."

This is the more sobering side of what we call mitigation. In mining, mitigation is the practice of extracting what we need while attempting to leave behind something better—or at least less harmed. Think of it as a kind of moral choreography: balancing the push of industrial ambition with the pull of environmental stewardship. But sometimes, as now, it involves the effort to prevent what happened before. Reminders of the Brumadinho collapse appear when you drive along the highway, hidden behind the trees: an open expanse of muddy ground, undulating in strange waves as if moving. But it did move once. It surged like a lethal surf when the dam came apart, sweeping away people, animals, and cars.

We travelled to Brazil every month, visiting a dozen mine sites to inspect more than twenty dams and facilities. We walked the perimeters, observing the work in progress, noting what looked sound and what raised questions. We watched as crews carefully excavated the dams, layer by layer, relocating tailings to sealed-off pits for long-term storage. Meetings with engineers, designers, and government officials followed, where presentations detailed the progress and the challenges ahead. Data from monitoring instruments was scrutinized, closure designs evaluated, and already decommissioned sites toured. The landscape we encountered was uncanny— canals and rolling green hills arranged with a precision just unnatural enough to be noticeable. It was the essence of the Anthropocene: a world reshaped by human hands. Yet, these engineered landscapes weren't just remnants of industry; they were a promise. Designed to withstand time and resist failure, they were built to be safer, stronger, and more enduring than even the natural ground beneath them.

Today, walking through the forest in search of a lookout to the valley below, I cross a section recently consumed by a low intensity fire. The bright, rainbow colouring of some of the eucalyptus tree trunks is covered with a dark, burnt husk, where flames had passed through. The ground is singed, but much of the upper canopy is spared. I come across a termite mound; an otherworldly pod rising from the forest floor. Such mounds are common across Brazil, dotting roads and highways. This one had borne the full force of the fire and appeared to have exploded from the inside out; its top split open and peeling back like the egg from *Alien*. I noticed remnants of the colony's intricate architecture scattered around it. I thought about the termites, how the heat and smoke must have signaled the coming disaster, yet they worked away, either oblivious or indifferent to the impending threat. As steam built up and pressure mounted, I imagined the last moments of the colony—a sudden and futile panic setting in far too late.

I thought of us—humanity—labouring on, even as the world burns, and the pressure rises around us. We have a line of sight greater than that of the termites. We have tools, intelligence, and the ability to think abstractly. We can learn from history, understand its context, and apply it to the present. Or we can keep our heads down, working blindly, never pausing to understand the systems that shape us or our place within them.

A couple of years ago, I was travelling through Colombia, helping with an estimate for the demolition and removal of a mine facility. The manager and I were talking about oil, its use and its extraction. In the middle of the conversation, he said to me, "Well, it really isn't that big a deal, because we have enough oil." Enough oil? Enough for what? This, for me, is the problem. We look metres ahead instead of kilometres. We plan for the moment, not for what comes next. What if the key to growth, exploration, or even survival, lies in something we don't yet know oil can do? Once we run out—whether in twenty years or three hundred—what's next? What replaces it? As we mine the earth for what we need, where is the long-term planning? Where is all this leading?

When I talk to students or young people distressed about the state of the world, I hear the same questions: "What should I study? Environmental science or political science? What should I protest? Who should I be angry at?" My answer almost always surprises.

You want to have an impact? Don't protest problems. Solve them. Study chemistry, the building blocks of everything. Study physics and engineering, the mechanics of how the world works. The biggest challenges we face—climate change, energy transitions, food security—aren't just political. They're material, scientific, technological. And here's what rarely gets talked about: there's already an army of brilliant people tackling these problems. They're working night shifts in labs, running experiments, publishing research, digging into the dirt—literally—to find solutions

Some of those people are here in Minas Gerais, with me. Where the Brumadinho dam once stood, a large conveyor belt crosses the highway like an overpass, built by the engineers and contractors tasked with the clean-up and remediation. At night, it looks like a rollercoaster in slow motion, humming as it carries away wreckage from the spill area. All that is left of the dam is a scar so vast it could be mistaken for a natural plain. There are people still missing, with families waiting for a call, years later. The wound is physical, emotional, cultural, and environmental. Like any wound, it needs tending.

Experts will study the damage, measure the impact, draft plans for rehabilitation. That's their job. Our job is different. We're here to stop the next catastrophe before it happens, before it leaves a scar even deeper than this one. Because in the end, we all share this place. Our lives, our needs, our ambitions, even our future, they all come from this rare earth.

ACKNOWLEDGEMENTS

A book like this crosses a strange threshold between decades of working in mines and decades of writing, and both worlds brought me the opportunity to engage with and learn from so many people.

My first home for the work that took me around the world was a company called Golder Associates. It was a company created and built by many brilliant people, and it was a place both my father and I called home for years. I need to thank Larry Lee and Masao Odaka for their years of patience and encouragement from the very start. Mark Pickering and Anthony Bottecia, for helping me learn fieldwork, how to stay safe, and for their humor and unwavering friendship. Dick and Jaquie Butler, for being our adopted family. Brian Wilson for the opportunities and the friendship. Mike Jefferies and Grant Bonin, for teaching me so much. John Lupo, for his willingness to share from his bottomless rucksack of wisdom, both in engineering and in life. Mickey Davachi, for the stories and the generosity of spirit. David Ritchie, for his patience and confidence in me.

My career would not exist without the support and encouragement of Terry Eldridge, who has set the bar for excellence in this field.

My brothers and sisters on projects near and far; I need to acknowledge the wonderful friendships of Vafa Rombough, Rob Chu, John Stuart, Karyn Gallant, Jay Diemert, Trent Collins, Carl Fitze, Zeke Baumgartner, and many others who got me through the night shifts, tough rotations, home safe, and made those years a joy.

Chris Anderson for a thousand conversations about work and life.

Don Hickson, for being as good a best friend as anyone could ask for, a brilliant engineer and an even better dinner buddy. From the Canadian Arctic to the streets of Lima, we have run the roads together.

My writing life has been a journey all its own, and along the way the encouragement of wonderful people at exactly the right time made all the difference. Chuck Bowie, whose friendship has been a constant. Al Horne, whose fierce mind and boundless spirit continue to inspire me; there is no teacher like him. Ayelet Tsabari, whose voice shows me what is possible. I owe a debt to the Summer Workshop at the Humber Writing Program, and David Bezmozgis for steering me to it. It was there where I met Colin McAdam, whose friendship and impact on my work will be lifelong.

Dillon Anthony, whose support comes daily. Brian Conoley, for his energy and wisdom. David Huebert, for the conversations and questions that became such a part of this book. Robin and Pete Saunders, in whose home so much of this book was written. Tom and Karen van der Pauw, who have been a source of love and encouragement since my teens.

Greg Harrison, for the chats, the music, and reminding me that creativity takes so many forms. Sacha Silva and Cristina Martinez, for their brilliance, kindness, and friendship.

Mark Anthony Jarman pushed me to write about my work and life from the day we met. It was an essay I wrote for him that I had in my hands when the book world came calling. For his support and endless encouragement, humor and friendship, I thank him.

In 2022, Carmine Starnino reached out to me about some non-fiction projects; that conversation became this book. Carmine's seemingly bottomless well of patience, courage, empathy, curiosity, and passion for this project is the reason this book exists. His resilience is the reason that our friendship still exists, and it is one that I treasure.

To Simon Dardick, Jennifer Varkonyi, and the entire team at Véhicule, thank you for the mountain of work and dedication to this project. There is no better team. To David Drummond, who created the most beautiful book cover I could have hoped for.

My sister Jennifer and my brother Dylan: you have been my models for courage, and I love you both.

Jennifer Baird, who I have adored since I was eleven years old, and who seems like she will tolerate me for a while longer. Your support and love make it all happen.

My parents did not live to see this book happen. My mother, Jane Anderson, brought art, beauty, and creativity to my life and made sure our house was one filled with books, music, and love.

My Dad, Bill Gilmer, opened the doors to the world that would become my professional life. His love and encouragement were what I carried with me to the far corners of the world. Several weeks after accepting the offer to write this book, my dad passed away from cancer. We did not know how much time we had left, but our last days were spent talking, hugging, and reminiscing about our lives as father and son.

He is present on every page.